Marine waters—p. 66
53-54

MONITORS

MONITORS

THE BIOLOGY OF VARANID LIZARDS

Dennis King and Brian Green

Illustrated by
Frank Knight, Keith Newgrain and Jo Eberhard

Krieger Publishing Company
Malabar, Florida

For Ruth and Iris

Original Edition 1993
University of New South Wales Press

Second Edition 1999
University of New South Wales Press Ltd
and
Krieger Publishing Company
Exclusive distributor for: Americas (North, Central, & South), Caribbean, Europe, and Africa.
1725 Krieger Drive
Malabar, FL 32950-3323 USA
Tel: (407) 724-9542 Fax: (407) 951-3671
info@krieger-pub.com

Copyright © D. King & B. Green

All rights reserved. No part of this book may be reproduced in any form or by any means, electronic or mechanical, including information storage and retrieval systems without permission in writing from the publisher.
No liability is assumed with respect to the information contained herein.

FROM A DECLARATION OF PRINCIPLES JOINTLY ADOPTED BY A COMMITTEE OF THE AMERICAN BAR ASSOCIATION AND A COMMITTEE OF PUBLISHERS:
This publication is designed to provide accurate and authoritative information in regard to the subject matter covered. It is sold with the understanding that the publisher is not engaged in rendering legal, accounting, or other professional service. If legal advice or other expert assistance is required, the services of a competent professional person should be sought.

Library of Congress Cataloguing-In-Publication Data:
A catalog record for this book is available from the Library of Congress, Washington, DC.

ISBN 1-57524-112-9

10 9 8 7 6 5 4 3 2

Printer Everbest Printing, Hong Kong

CONTENTS

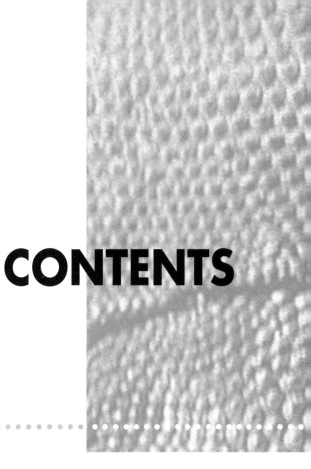

Preface **1**

Prologue **3**

CHAPTER 1 Introduction **5**

CHAPTER 2 Taxonomy and phylogeny:
Who's who, goannas' relationships and origins **9**

CHAPTER 3 Feeding **16**
- Diet **16**
- Food detection **20**
- Food handling **21**

CHAPTER 4 Breeding **25**
- Sex organs **25**
- Mating **27**
- Egg-laying **31**
- Clutch size **34**
- Hatching **35**
- Growth rates **37**
- Longevity **38**

CHAPTER 5 General behaviour **39**

    Home range **39**
    Activity areas **40**
    Activity patterns **42**
    Foraging strategies **43**
    Burrow use **44**
    Movement and body postures **45**

CHAPTER 6 Thermal biology: Regulation of body temperature **49**

    Burrows **49**
    Emergence time **51**
    Basking **52**
    Activity temperatures **53**
    Head–body temperature differences **56**
    Nasal tubes **58**
    Gular fluttering **58**
    Reflectivity of the skin **59**

CHAPTER 7 Respiration **61**

    Lungs **62**
    Heart **64**
    Blood physiology **64**

CHAPTER 8 Water use **66**

    Water losses **67**
    Salt-secreting glands **75**
    Water intake **77**

CHAPTER 9 Energy and food: The cost of living **81**

CHAPTER 10 Parasites **89**

    Ectoparasites **89**
    Endoparasites **92**

CHAPTER 11 Conservation and management **95**

    Australia **95**
    Asia **97**
    Africa **100**

Epilogue **102**

Suggested reading **103**

Index **112**

PREFACE

Why have we written a book about goannas — otherwise known as varanids or monitors? Apart from the obvious attraction of the unbounded wealth soon to befall us in royalties, we have long felt that a general text on the biology of these fascinating reptiles, for naturalists and students alike, was needed.

Goannas are a prominent Australian group of lizards. Of the 40 or so described varanid species in the world, more than 25 of them (including several more as yet undescribed species) are found in Australia. Most of the published information is contained in scientific journals, and many of these papers are written in a range of languages other than English. As such, this information is not readily available to, or easily understood by, the general public. We have therefore attempted to draw together and condense the more interesting aspects of varanid biology, and have tried to present this information in an easily readable form.

Monitors are different from most other lizards with respect to their ecology, behaviour, anatomy and physiology. They and some other lizards in the same lineage — called varanoids (helodermatids and lanthanotids) — plus a number of extinct groups are more closely related to snakes than are other lizards. Many of these differences have enabled them to flourish in harsh environments and for the larger species to serve in the role of top predators in many regions, especially in Australia. Many varanids are quite large; the Komodo dragon is the largest lizard in the world and has been reported to kill and devour humans. Their attributes have fostered great interest in the comparative anatomy and physiology of goannas, to the point where we know more about these aspects of their biology than about the general ecology of these lizards. Hopefully, this book will stimulate much needed research into their ecology and behaviour to help redress this imbalance.

Much of the information contained in this book is the work of other researchers. The list of suggested further reading at the end of this book indicates the sources of much of this information. We are particularly grateful for the useful conversations and correspondence we have had with Hans Otto Becker, Wolfgang Böhme, Mike Calver, Tristram Davies, Gil Dryden, Maren Gaulke, Hans-Georg Horn, Mike McKelvey, Gary Packard, Harvey Pough, Peggy Rismiller, Tom Tozer and Phil Withers.

Photographs were provided by Richard Braithwaite, Harry Butler, Graeme Chapman, Bernd Eidenmüller, Maren Gaulke, Hans-Georg Horn, R. J. (Hank) Jenkins, Ron Johnstone, Mike McKelvey and John Wombey.

The line drawings were prepared by Jo Eberhard, apart from Figures 1.1, 1.2, 2.3, 4.3, 4.4, 5.3, 5.4, 5.5, 5.6, and those in Table 9.2, which were prepared by Frank Knight. The graphs and figure annotations were prepared by Keith Newgrain.

Various drafts of the book were reviewed and criticised by Ruth King, Terry Dawson, Jo Eberhard, Allen Greer, Hugh Jones, Bob Sharrad, Rick Shine, Graham Thompson and Jill Wayment. They provided many helpful suggestions on ways to clarify and improve the book. However, the responsibility for any factual errors or gross grammatical faults that still remain must be accepted by Max King.

PROLOGUE

She moved into the small clearing and stopped in the spring sunlight. Head raised, tongue flickering, alert, she patiently searched the surrounding bush and waited. The only sounds were those from a gentle breeze rustling the gum leaves, and the distant chatter of parrots. Quietly, carefully, she started across the clearing, moving in a slow and undulating fashion. She examined and sniffed a small burrow, then turned and left it. On the far side of the clearing lay a rotting log — her next point of investigation. Again her tongue flickered, sampling the air for the scent of food. Strong claws tore at the log, dislodging a small skink from its shelter, but after a brief chase and a flash of sharp teeth it was seized. The skink struggled, but her teeth had pierced its body and the violent shaking of her head had ensured its sudden death. The skink's body was turned so that the head pointed towards the back of her jaws, and with a few twists of her neck it disappeared down her throat. All was still in the surrounding scrub. Her long, scaly body moved silently away and the hunt resumed. Spring had come to the island and with its arrival, the goanna had begun to hunt again.

INTRODUCTION

CHAPTER 1

What is a goanna? It is the common name in Australia for a lizard in the family Varanidae. However, in some parts of the country it is a name which is also given to any species of large lizard and is often incorrectly used for members of another family, the Scincidae. 'Goanna' is believed to be a corruption of the name 'iguana', which belongs to a different family of lizards found mainly in South America.

Goannas are distributed over a large part of the southern hemisphere, from Africa through southern Asia to Australia. In other countries they are commonly known as 'varanids' or 'monitor lizards'. The name 'goanna' is firmly entrenched in the terminology of Australians. No matter what scientists or others may call them, or how many kinds of lizards this common name might be applied to, it is a name which will continue to be used, even if only in Australia. The names 'goanna', 'monitor' and 'varanid' are used interchangeably in this book.

All varanids can be easily recognised as belonging to the same family of lizards. They are all insectivores or carnivores, although one species (*Varanus olivaceus*) eats large quantities of fruit for part of the year and snails and crabs during the rest of the year. All have a similar appearance, although they differ in size and body proportions. The body form of a typical monitor is shown in Figure 1.1.

THE BIOLOGY OF VARANID LIZARDS

Varanids have evolved into a number of species and have become specialised for some very different lifestyles. Some species have become adapted to a semi-aquatic mode of life, some are arboreal and others are adapted to living in some of the harshest deserts of the world. Others occur in mangrove swamps or dense rainforests, while a close relative, the earless monitor (*Lanthanotus borneensis*), has adopted a subterranean lifestyle in the jungles of Borneo.

An Australian monitor, *V. glauerti*, is exclusively arboreal in an area near Kakadu National Park in the Northern Territory, but on the Mitchell Plateau in Western Australia it occurs on rocky outcrops. *V. caudolineatus* is largely arboreal over much of its widespread distribution, but feeds mainly on terrestrial invertebrates. Several species with wide distributions (*V. gouldii, V. tristis, V. bengalensis, V. salvator*) utilise many different habitats and a variety of prey species, consequently their diets may differ greatly among regions.

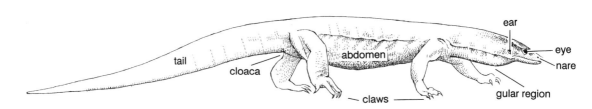

Figure 1.1 General morphology of a varanid

Varanids range in length from the pygmy species in Australia, with the total length of adults being from as little as 0.21 metres in *V. brevicauda*, to 3 metres for the Komodo dragon (*V. komodoensis*) of eastern Indonesia. An arboreal species in New Guinea, *V. salvadorii*, has been reported to reach a length of 4 metres, but this has not been verified. In weight they range from as low as 17 grams for adult *V. brevicauda* to at least 70 kilograms, and possibly over 100 kilograms for the Komodo dragon. An unconfirmed weight of 140 kilograms has been reported for a Komodo dragon which was shot in 1912, soon after the species first became known. The Komodo dragon is the largest lizard in the world today (Plate 2) but even it is dwarfed by the closely related but extinct species *Megalania prisca* (Figure 1.2) which was found over much of inland eastern Australia up to about 30,000 years ago. It reached a total length of 7 metres and is estimated to have weighed 650 kilograms. Early Aboriginal hunters may have encountered these awesome creatures while hunting in the Australian bush. The largest species found in Australia today (*V. giganteus*) is only about 2 metres long.

The closest living relatives of the Varanidae are believed to be the Lanthanotidae, or earless monitor, from Borneo, and the Helodermatidae, or beaded lizards, of Mexico and Central America.

INTRODUCTION

The lanthanotids are represented by a single species, the earless monitor (*Lanthanotus borneensis*) which has adopted a burrowing and apparently wholly subterranean existence. There are two species of helodermatids, the gila monster (*Heloderma suspectum*) and the beaded lizard (*Heloderma horridum*). They are slow-moving heavy-bodied animals which are highly specialised to locate nests and feed on eggs and young of mammals, birds and reptiles, and are the only species of lizards which are known to be venomous. These two groups of lizards and the varanids are all thought to be related to the extinct mosasaurs, an aquatic group of large reptiles which probably preyed upon fish and appear to be the group which gave rise to snakes.

Figure 1.2
Relative sizes of some varanids:
a. *Megalania prisca*
b. *V. komodoensis*
c. *V. rosenbergi*
d. *V. brevicauda*

The genus *Varanus* is most widespread in Australia, where about 30 of the world's approximately 40 to 50 species are found. There are 14 species in Asia, some of which also occur in Australia, and only 5 species in the whole of Africa. The world distribution of varanids extends as far east as the Solomon Islands, as far west as the west coast of Africa, and from the Caucasus in the north to the southern regions of Australia, with the exception of Tasmania (Figure 1.3).

The diversity and abundance of varanid lizards in Australia may be due to the absence of competition from mammalian carnivores during their evolutionary history, as compared with the abundant and diverse range of carnivorous mammals found in Asia and Africa. As many as ten different species of goannas may occur in the same general area in parts of Australia, but competition between them is minimised in several ways.

The Australian goanna that has been studied in most detail is Rosenberg's monitor (*V. rosenbergi*) (Plate 1). It can reach a total length of 1.5 metres and weigh up to 3.5 kilograms. It occurs along much of the south coast of Australia (Figure 1.4). *V. rosenbergi* have also been collected near Canberra and Sydney. This book deals mainly with the biology of this species and compares it with other species of varanids where appropriate.

Rosenberg's monitor is very abundant on Kangaroo Island off the coast of South Australia, at 35° 45' S, 138° 15' E, which has been separated from the mainland for 6000 to 8000 years. *V. rosenbergi* is a medium- to large-sized goanna and is the largest terrestrial predator native to Kangaroo Island. There are fossilised remains of larger marsupial carnivores at several sites on the island, but these species no longer occur there. Adult *V. rosenbergi* are now killed only by humans, vehicles and raptorial birds, such as wedge-tailed eagles.

Prior to 1957, *V. rosenbergi* was referred to as 'a melanistic form' of Gould's monitor (*V. gouldii*). It was then described as a distinct subspecies of *V. gouldii*. However, in a recent taxonomic revision of the varanid lizards found in Western Australia, it was recognised as being sufficiently distinct from *V. gouldii* to warrant the status of a full species, and subsequent taxonomic studies have provided further evidence supporting that decision.

Figure 1.3 World distribution of the genus *Varanus*

Kangaroo Island is almost as far from the equator as any location where varanid lizards can be found (Figure 1.3). The lace monitor (*V. varius*) occurs slightly further south in some areas of the Australian State of Victoria. The desert monitor (*V. griseus*) is the northernmost species of varanid. It occurs between 35° and 40° north of the equator in areas of the southern-most republics of the former USSR (Turkmenistan, Khazakstan, Uzbekistan and Tadzikistan).

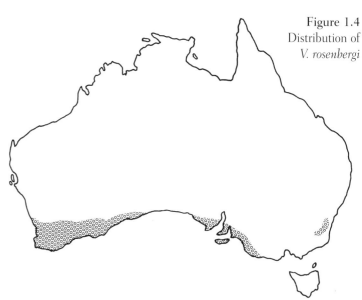

Figure 1.4 Distribution of *V. rosenbergi*

TAXONOMY AND PHYLOGENY

WHO'S WHO, GOANNAS' RELATIONSHIPS AND ORIGINS

Varanid-like reptiles first appear in the fossil record during the Cretaceous period, about 100 million years ago. After this period a wide range of both terrestrial and aquatic forms appeared. However, most of them became extinct by the end of the Palaeocene era (about 60 million years ago). These fossil varanoids are called platynotans and are found throughout Europe, Asia and North America.

The earliest fossil form that can definitely be ascribed to the Varanidae is *Telmasaurus*, found in Upper Cretaceous deposits in Mongolia. Another varanoid family, the Necrosauridae, was present in Europe from the Upper Palaeocene up to the Oligocene, between 65 and 26 million years before the present (BP). The genus *Varanus* first appeared in the fossil deposits of the Miocene (15 to 20 million years BP) in eastern Europe, Africa and Australia, and it is likely that the genus originated in Asia and radiated outwards into Africa and Australia. The nearest relatives of the varanoids (Varanidae, Lanthanotidae and the Helodermatidae) are the snakes and the mosasauroids, but they are sufficiently different from them to be placed in a sister group.

The first thorough attempt to classify the living species, all of which are currently placed in the genus *Varanus,* was that of Robert Mertens, who worked in Germany. His work, which was published in 1942, was severely hampered by the Second World War, since he did not have access to specimens in museums in many other countries. Mertens placed the 24 species of his original classification into eight subgenera, and some of the species were divided into several subspecies. In two subsequent studies he proposed two further subgenera, to give a total of ten. Mertens' classification was based almost entirely on the osteology (study of the bones) and the external morphology of the animals. It has since been modified extensively by workers who have used a range of techniques and who have also had access to larger collections of specimens from wider geographic areas. Most of the subgenera named by Mertens are now being revised and altered.

One technique used to investigate taxonomy and phylogeny is to examine and compare the number and morphology (shape and size) of the chromosomes. While the number of chromosomes in all varanids is constant with a diploid number of 40, their arrangement differs and the species can be placed in six groups on the basis of these characteristics (Figure 2.1). These groups are not consistent with the subgenera suggested by Mertens, as one of his groups contains members of four subgenera, and members of another subgenus belong to three different chromosome groups. This suggests the need for a new subgeneric division as proposed in Table 2.1.

Another way of studying the relationships among species is to compare the rates at which proteins extracted from body tissues move in a medium in response to an electric charge (electrophoresis). Such electrophoretic studies on blood, muscle and liver from varanids show relationships between the species similar to those found by chromosomal studies.

Another study has used an immunological method called micro-complement fixation (MC'F) to compare the structure of the protein albumin in different species. This method assumes that certain proteins mutate at a fixed rate and can therefore

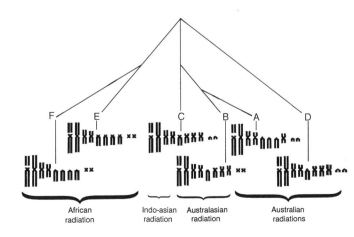

Figure 2.1 Chromosomal shape and relationships of varanids

Table 2.1 Proposed new sub-generic groupings of *Varanus* based on micro-complement fixation data

Species	Sub-genus
albigularis	*Polydaedalus*
exanthematicus	
niloticus	
yemenensis	
griseus	*Psammosaurus*
bengalensis	*Empagusia*
dumerilii	
flavescens	
indicus	*Euprepiosaurus*
jobiensis	
prasinus	
rudicollis	Un-named
salvator	
salvadorii	*Papusaurus*
olivaceus	*Philippinosaurus*
acanthurus	*Odatria*
baritji	
brevicauda	
caudolineatus	
eremius	
gilleni	
glauerti	
glebopalma	
kingorum	
mitchelli	
pilbarensis	
primordius	
scalaris	
semiremex	
storri	
timorensis	
tristis	
giganteus	*Varanus*
gouldii	
komodoensis	
mertensi	
panoptes	
rosenbergi	
spenceri	
varius	

be used as a molecular clock to estimate the degree of relatedness between species and when species diverged from each other. MC'F analysis of varanids has also produced similar results to the chromosomal studies.

These recent studies indicate that there are at least eight recognisable subgenera, although one technique suggests that an extra subgenus (*Philippinosaurus*) may exist, as was proposed by Mertens. Four of these subgenera are entirely different from those proposed by Mertens. The remainder are essentially the same, with deletions or additions of one or two species to two of them. The three unaltered subgenera (*Philippinosaurus*, *Papusaurus* and *Psammosaurus*) each consist of a single species. All those consisting of two or more species have been altered in some way to produce new groups of species (Table 2.1).

The phylogeny suggested by the MC'F study, which agrees very closely with the results of the chromosome morphology study, is shown in Figure 2.2. Such a phylogeny indicates that the family originated in southeast Asia and radiated east and southwards into Australia and westwards into the Indian subcontinent, the Middle East and Africa, but it was not able to conclusively determine which was the ancestral group. The suggestion of a southeast Asian origin has recently been supported by two gene-sequencing studies which have shown that, unlike agamid lizards which originated in Gondwana prior to its break-up, varanids originated in Laurasia about 65 million years ago and subsequently radiated into Africa and Australia during the late Tertiary period. All Australian species are in the same lineage and the small species in the subgenus *Odatria* appear to be derived from the larger Australian species. Based on the findings from MC'F and chromosomal studies the present day species seem to have appeared within the last 40 million years, which agrees with the fossil data from Australia and Africa. The common ancestry of species

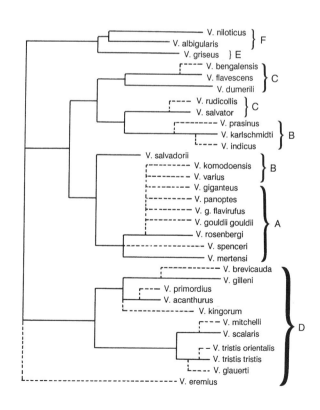

Figure 2.2 Proposed varanid phylogeny based on MC'F data

in Australia and Papua New Guinea, based on plate tectonics and the resulting connections between those two landmasses, may therefore go back as far as 30 to 40 million years.

There were several incursions by varanids into Australia from southeast Asia and Papua New Guinea during the past 40 million years, following the break-up of the Gondwana supercontinent. Between then and the Miocene (between 26 and 7 million years ago), Australia moved northwards and eventually came into close contact with southeast Asia. There has been considerable speciation in both of the major groups of varanids in Australia and some species of varanids have moved from Australia to New Guinea, or vice versa, in more recent times. Figure 2.3 shows the distributions of the proposed new subgenera.

There are major differences between varanid species in the structure of the male sex organs, which are known as hemipenes (see Breeding, page 25), and the bony structures within them called hemibaculae (Figure 2.4). The presence of similar but smaller organs (hemiclitoris) has now been determined in female varanids. Two separate studies of the male organs, one study of the female organs, and another comparing lung morphology, provide further support for revising the taxonomic groupings proposed by Mertens.

Figure 2.3
World distribution of proposed new varanid sub-genera (based on chromosome shape and MC'F data)

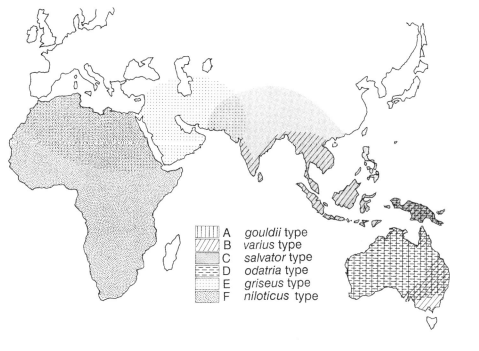

A *gouldii* type
B *varius* type
C *salvator* type
D *odatria* type
E *griseus* type
F *niloticus* type

While there is close agreement between recent studies in the species composition of the subgenera, they differ in their interpretation of the ways in which the subgenera relate to each other. For example, the inferred relationships between the Australian subgenus *Varanus* and three Asian subgenera (*Empagusia*, *Euprepiosaurus* and one unnamed) do not agree. Further studies using techniques such as gene-sequencing will hopefully fully resolve this issue.

The currently recognised species of *Varanus* are shown in Table 2.2. There is some confusion as to the appropriate names for some species, particularly in the *V. gouldii* group. In this book we use those currently accepted and used by most Australian herpetologists. Further taxonomic studies are certain to alter those used in Table 2.2. The names of some existing species will be changed and some newly recognised but as yet undescribed species will undoubtedly be described and named in the future (Plate 3). The present situation, where all species are placed in the single genus *Varanus*, will probably also change. Biochemical studies of some other reptiles have indicated the need to reclassify a number of groups and this has been carried out following re-examination of morphological characteristics.

Results of the chromosomal, biochemical and morphological studies mentioned above suggest that the varanids consist of several different genera, so the elevation of the proposed subgenera to full generic status should be seriously considered and investigated.

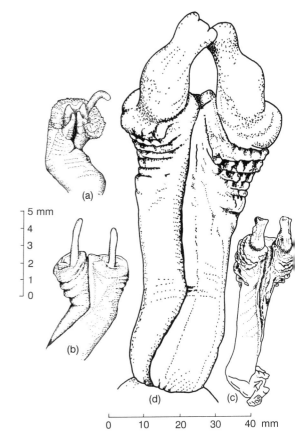

Figure 2.4
Hemipenes of some varanids:
a. *V. gilleni*
b. *V. primordius*
c. *V. gouldii*;
d. *V. komodoensis*
(after Böhme 1988)

Table 2.2 Currently recognised species of *Varanus* and probable changes in taxonomic status

Species	Distribution	To be changed	Probable change
albigularis	Africa		
exanthematicus	Africa		
niloticus	Africa	done	
ornatus	Africa	done	
yemenensis	Arabian Peninsula		
griseus	North Africa and Middle East		
bengalensis	Asia		
dumerilii	Asia		
flavescens	Asia		
komodoensis	Asia (Indonesia)		
olivaceus	Asia (Philippines)		
rudicollis	Asia		
salvator	Asia		yes
timorensis	Asia (Indonesia)		
indicus	Asia, New Guinea and Australia		
doreanus	New Guinea	done	
finschi	New Guinea	done	
jobiensis	New Guinea	done	
melinus	New Guinea	done	
prasinus	New Guinea and Australia		yes
spinulosus	New Guinea	done	
salvadorii	New Guinea		
yuwonoi	Asia (Indonesia)	done	
acanthurus	Australia		yes
baritji	Australia		
brevicauda	Australia		
caudolineatus	Australia		
eremius	Australia		
giganteus	Australia		
gilleni	Australia		
glauerti	Australia		
glebopalma	Australia		
gouldii	Australia		yes
kingorum	Australia		
mertensi	Australia		
mitchelli	Australia		
panoptes	Australia and New Guinea		yes
pilbarensis	Australia		
primordius	Australia		
rosenbergi	Australia		yes
scalaris	Australia and New Guinea		
semiremex	Australia		
similis	Australia and New Guinea		
spenceri	Australia		
storri	Australia		
tristis	Australia		yes
varius	Australia		yes

CHAPTER 3

FEEDING

DIET

The diet of lizards is studied either by identifying traces of prey left in faeces or by examining stomach contents. A high percentage of stomachs of many species of varanids contain no food, which indicates that prey capture is infrequent. As reptiles generally do not chew their food well, if at all, it can be readily identified. Stomach contents may be examined by dissecting dead, preserved museum specimens or by flushing the stomachs of live lizards to remove the food they have eaten. By either or both of these methods, the stomach contents of all known Australian varanid species have been examined.

All varanids which have been studied are wholly or largely carnivorous, although an arboreal species from the southern Philippines, *V. olivaceus*, is partly herbivorous. The invertebrate portion of its diet, which is mainly snails, varies from month to month but can reach up to 95 per cent of the total. *V. olivaceus* also feeds on a variety of fruits, particularly between May and September, and is the only goanna known to do so. Adult *V. olivaceus* eat approximately equal amounts of plant and animal foods over the course of a year, but juveniles eat much more animal than plant material.

FEEDING

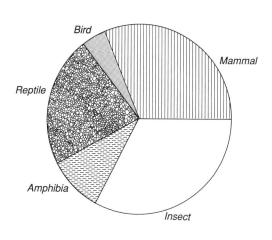

Figure 3.1
Diet of
V. rosenbergi
(% volume)

Australian pygmy monitors such as *V. acanthurus*, *V. baritji*, *V. brevicauda*, *V. glauerti* and *V. glebopalma* prey almost exclusively on invertebrates. The larger species, such as *V. giganteus*, *V. rosenbergi* and *V. varius*, which have larger gapes, also include large items in their diet, such as mammals, birds, reptiles and eggs. Very large prey items are not swallowed whole but are eaten in pieces sliced from the prey carcasses with their teeth. Only the larger species are powerful enough to utilise this source of food. They also feed on carrion from large prey. Most large species, however, still rely heavily on invertebrates for food.

When hunting for food, goannas walk with a characteristic swinging gait with their snout held close to the ground. The long forked tongue flicks in and out, transferring odours to sensory organs (Jacobson's organs). By this means, they can rapidly locate hidden prey, even if it is underground. Goannas use their strong claws as well as their snout for digging out prey.

The diet of *V. rosenbergi* on Kangaroo Island is shown in Figure 3.1. It contains a large number of invertebrates, but vertebrates form 65 per cent of the volume of their food. The majority of vertebrate food is probably scavenged from animals killed on the roads. Large species of goannas appear to eat whatever prey or carrion they encounter, as long as it is of suitable size. *V. rosenbergi* has a wide-ranging foraging pattern which includes many habitats, ranging from dense bush to open beaches. It often uses taste and smell when searching for hidden prey, and eats many small items of prey. This pattern of foraging is thought to resemble that of primitive varanids.

The diet of another large varanid from southern Australia, *V. varius*, which uses similar habitats to *V. rosenbergi*, has also been studied (Figure 3.2). Arthropods are represented to a similar extent as in the diet of *V. rosenbergi*, but mammals are more common in its diet and reptiles are eaten less often. *V. varius* eats food which is generally larger than that eaten by *V. rosenbergi*, but even the largest animals eat small prey when they encounter it.

Some smaller varanids, such as *V. gilleni*, have diets as varied as those of *V. rosenbergi* and *V. varius* in the content of small prey species (Figure 3.2), but they cannot eat large vertebrates because of their small size. Their diet thus consists almost entirely of invertebrates.

However, adults of the largest species, *V. komodoensis* rely more on a sit-and-wait hunting strategy than on a wide-foraging one and have fewer invertebrates in their diet. Whereas invertebrates comprise most of the diet of many species of varanids, they presumably provide insufficient energy to warrant the emergence of *V. komodoensis* from their place of ambush. Large Komodo dragons hunt on the ground, and often wait beside game trails. They immobilise passing prey, including Asian buffalo, by seizing them by the leg and tearing their hamstring tendons. They can then kill the crippled prey by tearing at the throat and belly, as they do with their usual smaller prey, such as pigs and deer. Carrion also forms a substantial part of their diet, as it does in *V. bengalensis* and *V. salvator*. Small Komodo dragons usually hunt in the trees and feed mainly on small lizards (skinks and geckoes) and invertebrates, while medium-sized animals feed predominantly on birds and rodents. Hatchlings of most other large species of varanids, including *V. bengalensis* and *V. varius*, feed mostly or totally on invertebrates.

Ambushing is not a typical hunting strategy for varanids, but it does occur in *V. komodoensis* and may also occur in some other species, such as *V. acanthurus, V. brevicauda, V. glebopalma* and *V. tristis*.

Eggs of other reptiles (turtles, crocodiles, lizards) or of birds may sometimes form a major portion of the diets of several species of goannas, including *V. giganteus, V. tristis* and *V. varius* in Australia, *V. indicus* on Guam, *V. salvator* in Asia and *V. niloticus* in Africa.

Goannas forage widely and their attention is often engaged in activities such as digging for prey. They thereby expose themselves to predation more than would an ambush predator. Smaller goannas are particularly vulnerable.

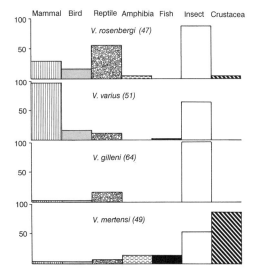

Figure 3.2
Diet of four varanid species (% occurrence)

Cannibalism occurs in several varanid species (*V. rosenbergi, V. gouldii, V. giganteus, V. komodoensis* and *V. bengalensis*). Occasionally it involves feeding on carrion, but predation by wild goannas on live members of their own species has been observed in *V. komodoensis, V. griseus, V. gouldii* and *V. giganteus*.

The scope of the diet is influenced by season, the location of the animal and by prey diversity and abundance. As most of the energy used by varanids in hunting is spent locating prey, rather than in subduing it, a wide range of prey items is acceptable to goannas. Varanids opportunistically use locally abundant types of prey. Should a small item such as a beetle or a cockroach emerge from a burrow or from under a log, it will be eaten as readily as larger prey such as a lizard or small mammal. The composition of the diets of *V. gouldii* from four different study sites varied markedly, as did those of *V. bengalensis*, *V. mertensi*, *V. panoptes*, *V. giganteus* (plate 6), *V. tristis* and *V. caudolineatus* from different locations. A major food item of *V. albigularis* in Namibia is snails, which are also eaten often by *V. niloticus*, *V. griseus* and *V. salvator*.

Some varanids, including the Australian species *V. mertensi*, are highly semi-aquatic, mainly capturing and feeding on fish, crustaceans or molluscs in the water (Figure 3.2). *V. mertensi* (Plate 5) swim and also walk underwater along river bottoms searching for prey. However, not all the species commonly found near water eat aquatic prey. The diet of *V. indicus* is predominantly lizards, crabs, fish, grasshoppers and small mammals. The most important food of *V. salvator* in some areas is grasshoppers and crickets, while elsewhere it mainly feeds on crabs, small rodents and any other prey it can capture. Other varanids (for example, *V. tristis* and *V. varius*) that are partly or completely arboreal include eggs or nestlings of tree-nesting birds in their diet, but several arboreal species (*V. tristis*, *V. caudolineatus*, *V. timorensis*, *V. scalaris*) also forage on the ground.

Most varanid species forage on the ground and pay particular attention to potential hiding places such as burrows or surface debris under which prey may shelter. Varanids such as *V. eremius*, *V. tristis*, *V. varius*, *V. griseus*, *V. glauerti*, *V. albigularis*, *V. gouldii*, *V. giganteus*, *V. komodoensis* and *V. bengalensis* apparently learn to recognise good locations for finding food or other important resources and regularly return to them during their foraging. These locations include resting and nesting areas of many types of prey. This can lead to dietary specialisations among some monitors, as can the high abundance of certain prey types in their particular habitats.

Some aspects of the anatomy, physiology and behaviour of varanids differ from those of other lizards and allow them to forage differently. These include the structure of some organs, high body temperatures and efficient respiratory systems which enable them to maintain high activity levels. These traits enable the largest species to pursue, capture and eat large vertebrates, but most small or medium-sized species generally feed on invertebrates. When it is available, large amounts of food can be eaten in a very short period. Several species of *Varanus* have been observed to eat 15 to 30 per cent of their own body weight at one time, and *V. komodoensis* has been reported to eat over 60 per cent of its own weight in one meal. The varanids are probably descended from medium-sized lizards which fed predominantly on insects which they encountered as they foraged widely throughout their surroundings. Present day varanids may often feed on invertebrates because they encounter them more frequently than vertebrates as they forage. In addition, the smaller species may feed almost entirely on invertebrates because they are not powerful enough to subdue most vertebrates.

FOOD DETECTION

THE SNOUT AND NASAL SACS

The snout of a varanid is important in hunting for prey. Each external nostril (nare) opens into a part of the skull called the nasal capsule, which consists of a number of chambers (Figure 3.3). The nasal capsule in varanids is narrow, elongated and cartilaginous. In the anterior region it is covered only by skin and connective tissue, whereas in other reptiles it is covered by bone. The roof of the posterior region is continuous cartilage in varanids.

The nasal tubes lead from the nares to the anterior chambers, which are not sensory organs, being lined with flattened, non-sensory cells. Spongy tissue surrounds the nasal tubes and the anterior chambers and provides a mechanism for closing off the nasal passages to protect the olfactory chambers. The olfactory chamber is less extensive in varanids than in most other lizards. It is lined with sensory olfactory cells. The olfactory chambers connect with the mouth via internal nares (choanae).

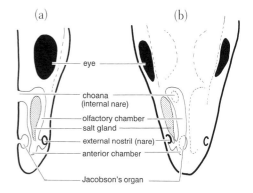

Figure 3.3
Morphology of the snout region:
a. side view
b. top view

The position of the nares on the snout varies between species. This may be of adaptive significance for goannas which occupy different types of environments. It has been suggested that those species that have nares close to the eyes may forage through litter which they disturb with their snouts. The nares of *V. mertensi* are located on the top of the snout, which aids in their utilisation of an aquatic environment.

JACOBSON'S ORGANS

In addition to the olfactory chambers, lizards and snakes possess a pair of special sense organs called Jacobson's organs, which are also located in the nasal capsules above the roof of the mouth (Figure 3.3). Their large size in the Varanidae suggests that they play an important part in the sensory capabilities of goannas. These organs are connected to the roof of the mouth and the nasal sacs by ducts. Jacobson's organs enable a lizard or snake to detect scent particles which have been collected and transferred to the roof of the mouth by the tongue. Jacobson's organs contain sensory cells similar to those found in the olfactory chambers. These sensory cells are connected to the accessory olfactory bulb of the brain by the vomeronasal nerves. Jacobson's organs supplement the olfactory chambers as organs of smell, and are also important in prey location and social interactions.

FOOD HANDLING

THE TONGUE AND HYOID APPARATUS

In its primitive form the main role of the tongue in reptiles was the manipulation of food. In most living species of lizards and snakes the tongue is used to manipulate food in the mouth during feeding and also to convey scent particles to the Jacobson's organs. Those species which are able to protrude the tongue further from the mouth than more primitive lizards have more complex Jacobson's organs. In varanids, however, the tongue is modified to fill a sensory role for use in locating prey and in social behaviour. The Varanidae are the only family of lizards which uses the tongue exclusively for sensory functions. Snakes' tongues are similar in appearance and are also used only for sensory purposes. This, and the fact that many snakes also utilise an active foraging pattern to locate prey, has led to suggestions of a possible close evolutionary relationship between varanids and snakes. However, the musculature of their tongues differs substantially, even though snakes are now known to be more closely related to varanids than to other lizards.

The structure of the tongue in varanids is very different from that of other lizards. It is much longer and narrower, with a deep incision in its tip. It does not move freely in the mouth but fits into a sheath when it is retracted (Figure 3.4). It does not have the roughened dorsal surface found in other lizards. It consists of four parts: a long posterior portion with longitudinal muscle fibres; an area of mainly circular fibres; the main body of the tongue, which has a very complex set of muscles; and the incised tips, which also have muscles running in various directions.

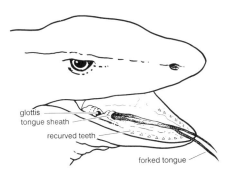

Figure 3.4
Mouth and tongue of a varanid

The tongue is partially supported by a complicated structure of bone and cartilage called the hyoid apparatus (Figure 3.5). The hyoid apparatus of varanids is similar to that of lizards from other families but it has some specialisations, particularly to its anterior portions. There is greater mobility between the joints, and the lingual process enters the connective tissue below the tongue rather than the body of the tongue itself. The superficial throat muscles are exceptionally broad. The hyoid apparatus is relatively heavier and more elongated than in other lizards. This specialisation of the hyoid apparatus, its mobile joints, and associated throat muscles are used to force food into the throat. This has enabled the function of the tongue to change from a manipulatory to a sensory role.

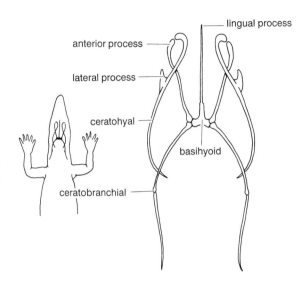

Figure 3.5 Hyoid apparatus (after Smith 1986)

In its sensory role, the tongue is protruded and flickered rapidly. The tips move independently, thereby increasing their sampling area (Plate 7). Varanids are the only family of lizards which lack taste buds on the tongue, and are similar to snakes in this respect.

Particles collected from the air by the tongue are transported to the roof of the mouth close to the openings of the Jacobson's organs. The deeply incised tongue does not insert particles directly into the Jacobson's organs. The function of the long tips is not known but may be to increase their surface area. The tongue is not flattened when it is protruded. It is a cylindrical mass of muscle with a constant volume. When protruded, it becomes longer and thinner. When the tongues of goannas are extended, their length may increase by 72 to 90 per cent. This is much greater than for lizards of other families. Long narrow tongues can be protruded further than shorter, stout ones, and therefore are able to sample a larger volume of air and collect more scent particles than shorter ones. The flickering tongue is highly characteristic of varanids as they search for food, and may have been very important in determining the evolution of their foraging behaviour. Many other families of lizards are unable to utilise a similar strategy of foraging because their tongues lack the length and flexibility of varanid tongues. The flickering tongue enables varanids to locate hidden prey and to track down injured prey which may have escaped from an attack. A well-developed ability for chemical discrimination is present at hatching in varanids, and is highly beneficial in foraging, reproduction and communication with other varanids.

In most animals, food is transported through the mouth by regular cycles of movement of the tongue, in conjunction with associated jaw movement. Varanids, however, do not use the tongue in the movement of food. Instead, inertial feeding is used, in which the prey is thrown backwards in the mouth and the jaws move rapidly forwards to surround it. Two or three such thrusts may be required to deal with small items of prey while five to twelve may be needed for larger prey. The tongue is protected in its sheath on the floor of the mouth during these movements. Once the food is well within the gular region (soft tissues of the throat; Figure 1.1, page 6), the hyoid apparatus moves it back into the oesophagus. Meanwhile, the neck is bent from side to side to assist in moving the prey down the throat. Large prey are often pressed or hit against other objects to force them into the throat. In captivity, the sides of cages are often used for this purpose.

The tongue is probably not used in transporting liquid when varanids drink. They insert their snout into the liquid and may immerse it entirely. Drinking is probably accomplished by suction created by pumping of the hyoid apparatus and the muscles attached to it. Once liquid reaches the throat, it is forced further back by lifting the head and compressing the hyoid region. This behaviour has been observed in the wild in *V. komodoensis, V. gouldii* and *V. panoptes* and in several other species in captivity.

TEETH AND JAWS

The teeth of varanids are large, sharp, and recurved. They are laterally compressed with anterior and posterior cutting edges and are arranged in long tooth rows (Figure 3.4). Recurved teeth have greater prey-holding ability than straight teeth and are useful in catching and successfully handling prey. The teeth are normally hidden in a fold in the gums and are not readily seen. However, their presence is obvious when the animals bite, as they can cause deep lacerations. Handling large prey places great strains on teeth and they often break off where they are attached to the side of the jaw. Replacement teeth develop from the pulp cavity and become fused to the outside of the jaw after bone at the site of the old tooth is resorbed.

The jaw and throat musculature in the larger species of varanids is adapted to feeding on large prey. The jaws are able to close rapidly and are thus suited to the capture of fast-moving prey. After capture, prey must be subdued and correctly orientated before being swallowed. Large items may be given a number of bites, be shaken in the jaws or pressed against objects in the process of being suffocated. Larger prey are smashed against the ground or other solid surfaces. The prey are held securely until all movement ceases. Small prey are subsequently swallowed whole, head first. The larger species of goannas dismember especially large prey. They hold the prey item with their forefeet and tear it into pieces which they can swallow with jerking movements of their heads. In this way, a 50 kilogram Komodo dragon (Plate 4) has been seen to dismember and completely devour a 31 kilogram pig carcass in 17 minutes.

In most lizards, the upper jaw can move independently of the base of the skull. This process is called cranial kinesis and is important in assisting the movement of prey into and within the mouth. Compared with its development in other lizards, kinesis is highly developed in varanids. It aligns the upper and lower tooth rows so that prey may be more firmly gripped. This is important when capturing and holding the small elusive prey upon which most species of goannas predominantly feed, and which they are very adept at seizing. Kinesis also allows the mouth to open wider when larger items of prey are being eaten. The part it plays in inertial feeding is also important. When the head is jerked sharply backwards, the jaws move forwards and thus gain a grip further along the body of the prey. The hyoid apparatus (Figure 3.5) then helps to force the prey into the throat.

From the mouth the food is forced along the oesophagus by the hyoid apparatus and into the elongated, muscular stomach (Figure 3.6) which serves mainly as a storage organ. Beyond the stomach, the pylorus leads into the small intestine (ileum) where most digestion occurs. The ileum in turn leads into the rectum, which is separated from the cloacal chambers by a sphincter muscle. The gut of varanids and other carnivorous reptiles is short compared with those of herbivorous lizards, as digestion is a comparatively simple process in carnivores.

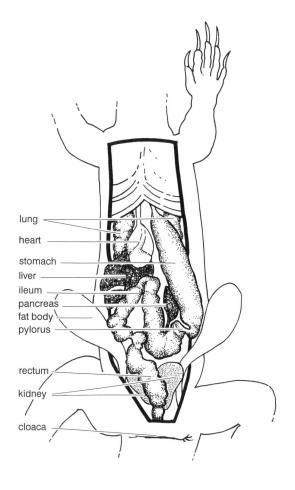

Figure 3.6
Internal organs of
V. rosenbergi

BREEDING

CHAPTER 4

SEX ORGANS

Male goannas are distinguished by their generally larger size and the presence of two lateral bulges at the base of the tail, although these are not obvious in some species. Each bulge contains a hemipenis, the copulatory organ of the male. Each hemipenis is everted from the cloacal opening whenever the animal is aroused, either for copulation or in response to threat. The hemipenes are erected and retracted by means of muscular contractions and variations in blood supply. These organs are very elaborate structures (Figure 2.4, page 14). The flamboyant appearance of the hemipenes is probably important during courtship display or mating. The hemipenes of varanids each have two lobes, and each lobe bears a terminal cartilage or bone, called the hemibaculum.

Female Rosenberg's goannas also have organs that resemble hemipenes and, although smaller and less developed than those of males, they can often make it very difficult for humans to determine the sex of this species, particularly when comparing females with juvenile males.

The testes and ovaries of varanids are located within the abdomen (Figure 4.1) and are relatively small when the animals emerge from the general inactivity of winter. During spring the gonads of both sexes increase in size (Figure 4.2) and gametes (sperm and ova) begin to develop. The testes reach their maximum size in December and at about the same time the ovaries shed ova into the oviducts. Over several weeks the ova increase substantially in size as large quantities of creamy coloured yolk are added to them. By mid-summer Rosenberg's goannas are ready to mate.

Some species (mainly the smaller species but also including *V. salvator*) are able to breed at an early age. Male *V. brevicauda* can breed when 10 months old and females in their second year. *V. komodoensis* begin to breed when they are three to five years old and weigh about 20 kilograms, and *V. albigularis* probably become reproductively active when they are five to six years old.

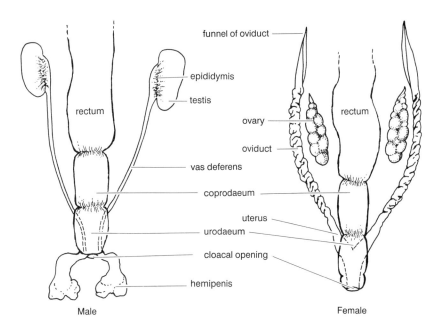

Figure 4.1 Sex organs of a male and a female varanid

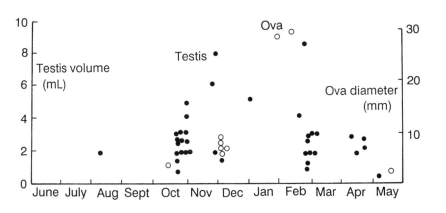

Figure 4.2 Seasonal changes in size of testes and ova in *V. rosenbergi*

MATING

Mating in *V. rosenbergi* is a surprisingly tender event. Once a male goanna has located the burrow of a female that is in breeding condition, he constructs a burrow of his own a few metres away. Over several days the male and female spend an increasing amount of time in each other's company. Eventually the pair copulate. Copulation is preceded by the male entering the burrow of the female, leaving his tail protruding from the burrow entrance. After a few minutes he backs slowly out of the burrow, leading the female, head first, to the surface. As he leads her, he nuzzles and licks her neck, and when they are both in the open outside the burrow he frequently rubs the side of his head along her back. He also nuzzles her cloacal opening before moving his body forward across her back, until their cloacas are side by side. He may approach her from either side as he has paired copulatory organs. The male uses his tail to raise that of the female off the ground, while the female raises her hindleg closest to him, allowing him to insert the adjacent hemipenis. Insertion is accompanied by a few lateral pelvic thrusts after which the pair remain coupled and still for about ten minutes (Figure 4.3). The female, who appears extremely nervous at this time, then breaks free and retires to her burrow.

The male remains on the surface, basking on the entrance mound and occasionally digging and scraping further soil onto it. After a further ten minutes or so, the male re-enters the female's burrow and persuades her to re-emerge for further copulations. This activity goes on for several days, with the male using his hemipenes alternately. During this period of intense breeding activity the pair may come to share the same burrow. After some days, the intensity of copulation declines and finally only occupies a few hours in the morning. After this, the goannas separate and forage independently.

Figure 4.3
Copulation in
V. rosenbergi

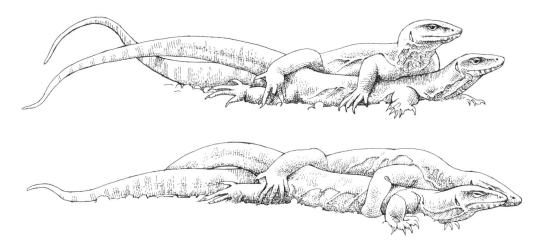

If another male enters the immediate vicinity of the breeding pair, the couple separate and the female retreats to her burrow while the male chases off the intruder.

Good data on breeding in the wild is only available for a few large *Varanus* species (*V. albigularis*, *V. komodoensis*, *V. olivaceus*, *V. bengalensis*, *V. varius* and *V. rosenbergi*). The large African species, *V. albigularis* (males weigh up to 8 kilograms, females up to 6.5 kilograms) has a rather different form of mating behaviour. In Namibia, adults (five to six years and over) breed during the cool, dry season (July to August), when vegetation and invertebrate numbers increase greatly. The males then begin to seek receptive females, which are located in trees (or sometimes in burrows) in which they have been sheltering for several months. The males travel over very large distances (up to 4 kilometres a day) in search of them. At the end of the mating season the males also retreat to refuge areas for 2 to 3 months. The long distances travelled are a result of the promiscuous mating behaviour of the males who mate with a number of females, to whom they are attracted by pheromones (attractive odours) produced by the females.

The energy needed by the females for producing eggs is stored as fat during their main feeding period, and is conserved by the female remaining within a very small area (less than 100 square metres) during the breeding season. Their limited mobility also makes it easier for the males to locate them. The only time when both sexes move extensively is during the wet season (January to April), when their major food source (land snails) is abundant.

Female *V. albigularis* lay their eggs in the period from September to December, about 35 days after mating, and most hatchlings appear after 135 to 150 days of incubation in the hot–wet season (February to March) when food is abundant. Eggs incubated at higher temperatures (30°C to 31°C) hatch earlier, but produce lighter and smaller young than do eggs incubated at temperatures 3°C to 4°C lower, and larger hatchlings tend to survive better than smaller ones. Hatchlings of *V. albigularis* are able to use chemical cues to determine whether snake species are venomous or not, which gives them a survival advantage.

Females of the Australian species *V. tristis* also remain in trees for long periods and mate there, often with several males. It is likely that these males, like *V. albigularis*, locate receptive females during the mating season by the detection of pheromones.

V. griseus in Uzbekistan hibernate from October to April and mate during a very short period (10 days) in May, when the males locate females by following their scent trails over distances greater than 1 kilometre, and from which they can determine the sexual condition of females. After the mating season, *V. griseus* are unsociable until the start of hibernation, despite the fact that the core areas of their home ranges broadly overlap. Apart from the breeding season, there is little interaction between individuals in populations of *V. albigularis, V. rosenbergi* and *V. tristis.*

The mating behaviour of species other than *V. rosenbergi* has been observed in the wild (*V. albigularis, V. komodoensis* and *V. varius*) and in captivity (*V. varius, V. bengalensis, V. olivaceus, V. timorensis* and *V. gilleni*). Behaviour of all species is generally similar, but there are some characteristic differences between species. Scratching, nuzzling the head and neck area and licking are important in all species. In some (*V. komodoensis, V. olivaceus* and *V. gilleni*) the male occasionally bites the female while he lies on her back, but this was not seen in *V. timorensis* or *V. rosenbergi. V. bengalensis* females sometimes bite the male during mating. Male *V. komodoensis* and *V. bengalensis* usually immobilise the female during mating by holding her in a tight embrace with his front legs. Females of these species are aggressive during mating, and could damage the males by biting or clawing them. This immobilisation has not been observed in *V. varius, V. rosenbergi* or *V. olivaceus*. In *V. olivaceus, V. timorensis* and *V. varius*, the males use either hindleg to lift the base of the female's tail prior to inserting the hemipenis. The mating behaviour of male varanids of different species has to be flexible so they can react to the varied responses of the females. Varanids generally copulate frequently over a period of several days. One pair of wild *V. varius* was seen to copulate 16 times in three hours.

Some species (*V. komodoensis, V. bengalensis* and *V. rosenbergi*) seem to form pair bonds of varying duration. In captivity, mating generally occurs only after both animals have been kept in the same cage for a lengthy period. However, wild female *V. varius* have been seen mating with several males within a few days. Mating assemblages, which usually consist of one female and several males, have been seen in *V. varius* and *V. panoptes*. The largest male *V. varius* eventually chases away other males that approach the female.

Males in at least 19 species of *Varanus* engage in what is called ritual combat. The larger species (*V. komodoensis, V. bengalensis, V. olivaceus, V. salvator, V. indicus, V. niloticus, V. dumerilii, V. albigularis, V. panoptes, V. spenceri, V. varius, V. gouldii* and *V. mertensi* (plate 21)) initially wrestle one another in an upright stance and clutch each other with the forelegs only (Figure 4.4a), whereas the smaller species (*V. gilleni, V. semiremex, V. similis, V. caudolineatus* and *V. timorensis*) clutch one another with both the forelegs and the hindlegs during the early stages of these encounters. Such fights can also involve the antagonists rolling

Figure 4.4
Ritual combat postures in:
a. *V. komodoensis*
b. *V. gilleni*
(species not drawn to same scale)

along the ground while entwined (Figure 4.4b). This may have a phylogenetic derivation as these groups of species all correspond with recognised clades (Chapter 2, Taxonomy and Phylogeny). During ritual combat varanids try to overpower one other, and severe wounds are occasionally inflicted by the teeth and claws in some species. Usually it is only after dominance is determined that the subordinate animal is bitten by the winner. Some species (*V. bengalensis* and *V. salvator*) apparently do not bite one another during this combat, while other species do (*V. gilleni, V. niloticus, V. olivaceus, V. komodoensis* and *V. varius*).

Similar behaviour has recently been observed in two captive *V. varius*, one of which was a female, so this behaviour may also serve other purposes, such as rivalry for food. Combat between similar-sized wild females has been observed in *V. salvator cumingi* near an animal carcass in the Philippines.

EGG-LAYING

Although egg-laying in some species of varanids may occur several times during a period of several days, only *V. salvator* is known to breed more than once a year in the wild. Other species have done so in captivity, perhaps due to an unnaturally abundant supply of food.

Varanids from the temperate regions lay their eggs during late spring and summer (October to February), whereas tropical species lay their eggs at different stages of the wet season (April to June, September to February). It has been shown by studies of *V. gouldii* and *V. tristis* that the time of egg-laying in populations of the same species in different locations may vary substantially.

The time between fertilisation and egg-laying of *V. rosenbergi* is not known, but during this period the leathery shells that are characteristic of many reptiles are formed in the lower portion of the oviducts. In captivity, this process takes between 4 and 6 weeks in other species of monitors. After this process occurs, the female *V. rosenbergi* locates an active termite mound (usually *Nasutitermes exitiosus*) and begins to excavate an access hole at the top, or high on one side of the mound (Plate 8). She digs a tunnel approximately 50 to 60 centimetres deep, towards the centre of the termitarium. At the end of this shaft she makes a large cavity (Figure 4.5). The digging of this egg chamber is usually completed within a day, but can take longer if the termite mound is very hard. On completing the excavation the female sits in the access shaft and begins to lay her clutch of eggs into the deep chamber, with only her head protruding from the mound. Usually 10 to 17 eggs are laid, weighing approximately 26 grams each and measuring about 50 x 30 millimetres. Most species of varanids lay their eggs (oviposit) during the daytime, but on some occasions *V. rosenbergi* have been seen ovipositing well after the onset of darkness. Captive specimens of five other Australian species of varanids (*V. giganteus, V. indicus, V. mertensi, V. prasinus* and *V. varius*) and two African species (*V. albigularis* and *V. exanthematicus*) have also been observed to oviposit nocturnally.

The eggs of varanids differ greatly in size. The average weight of those of *V. gilleni* is only 4 grams each and they measure 16 x 27 millimetres, while those of *V. komodoensis* weigh up to 125 grams each and measure 55 x 85 millimetres. Varanid eggs are extremely rich in lipids (fats) that provide the main energy source for the developing embryo. The general composition of goanna eggs is shown in Table 4.1.

Egg-laying by *V. rosenbergi* occupies a few hours, after which the female begins to refill the excavation. She pushes soil up the mound using her hindlegs, scratching surface mound material towards the hole with her forelimbs. The female takes several rests during the refilling process. She often moves around the summit region of the mound, apparently marking it with scent from her cloacal region. She may also stop to bask.

The male may associate with the female as she labours to completely close off the egg chamber, but he performs only occasional desultory scratching at the mound before returning to his foraging. Both adults pay regular visits to the termitarium in the days immediately after egg-laying. This probably prevents other goannas from attempting to deposit their eggs in the same mound. It may also serve to ward off any diurnal predators trying to rob the nest chamber before the termites have reconstructed the mound. During this post-mating period the male may still attempt to court and mount the female. She rejects these attempts by adopting a raised aggressive

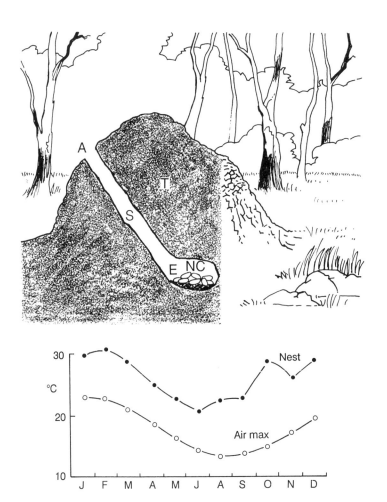

Figure 4.5
Goanna nest in a termite mound
Legend:
termite mound (T)
access hole (A)
shaft (S)
nest chamber (NC)
eggs (E)

Figure 4.6
Monthly changes in nest and air temperature

Table 4.1 Composition of eggs of *Varanus rosenbergi*

Egg	Weight (g)	Dry matter (%)	Water (%)	Shell (%)	Fat (%)	Protein (%)	Energy (kJ/g)
1	25.7	32.0	68.0	2.6	13.4	11.6	8.2
2	24.9	33.3	66.7	2.9	14.3	11.7	8.6
3	26.4	28.9	71.1	2.9	13.8	9.0	7.7
Mean	25.7	31.4	68.6	2.8	13.8	10.8	8.1

stance. Egg-laying is generally completed by late summer in *V. rosenbergi*, after which mating pairs separate. The gonads decline in size at the end of the breeding season (Figure 4.2, page 26).

The termites reconstruct the mound around the *V. rosenbergi* eggs so that they become totally encapsulated. Termitaria are excellent incubation chambers for reptile eggs. The termites regulate the temperature of the mound to around 30°C throughout most of the year, apart from a slight cooling in winter. The humidity in the mound is always near saturation. Therefore, the goanna eggs and embryos are maintained at a suitable temperature and humidity during their development (Figure 4.6). The eggs are able to absorb moisture from their surroundings and increase in weight as they develop.

Although not all species of goannas lay their eggs in termitaria, at least some of the larger species are known to do so (*V. albigularis*, *V. bengalensis*, *V. griseus*, *V. salvator*, *V. giganteus*, *V. gouldii*, *V. niloticus*, *V. rosenbergi* and *V. varius*). *V. varius* and *V. prasinus* also use termitaria in trees. The surfaces of some termite mounds are very hard, and only large species with sharp and powerful claws can open the mounds and use them for nesting places. Termitaria provide protection for the eggs from predators, and the young of some species (*V. niloticus*, *V. varius*, *V. rosenbergi*) sometimes remain within them for up to several months after hatching. Some species (*V. panoptes*, *V. giganteus*, *V. tristis* and *V. bengalensis*) bury their eggs deep in the soil, especially along the margins of creek beds. Also, some varanids appear to lay their eggs in communal nests that give the appearance of a warren (*V. komodoensis*, *V. bengalensis* and *V. panoptes*). The locations of the nests of most small species are not known, but *V. tristis* have been observed laying eggs in burrows dug at the base of trees.

CLUTCH SIZE

The clutch size of varanids is generally related to their body size (Figures 4.7 and 4.8). However, there are some exceptions. Some species, such as the desert-dwelling *V. tristis* produce uncharacteristically large clutches of small eggs. The maximum clutch size reported in varanids is 51 eggs in *V. salvator*.

Fully developed eggs occupy all the available space in the body cavity of female varanids and the animal becomes highly distended. In at least two species (*V. tristis* and *V. scalaris*) most of the gravid females carrying nearly developed eggs have empty stomachs, probably because of the limited abdominal space remaining and the pressure exerted on the internal organs by the eggs. Feeding probably stops during the late stages of egg development in most varanid species. Female *V. albigularis* are reported to be largely inactive for approximately one month prior to egg-laying, perhaps as a consequence of being less mobile and thus more prone to predation.

Data obtained from eggs of captive animals which have been incubated at similar, constant temperatures, show that incubation time is related to egg size or body size. The eggs of large species are larger than those of smaller species and take longer to hatch. At 28.5°C to 30.0°C incubation periods range from 77 days for *V. brevicauda* to 250 days for *V. salvator* (Figure 4.9). Incubation periods under constant temperatures in captivity are shorter than incubation periods observed in the wild. This may be due to

Figure 4.7 Clutch size and body size in *V. acanthurus*

Figure 4.8 Clutch size and body size in varanids

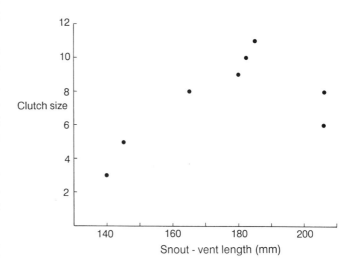

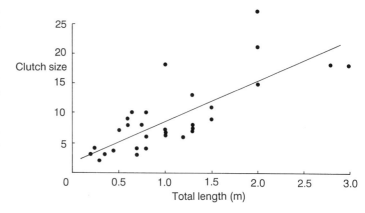

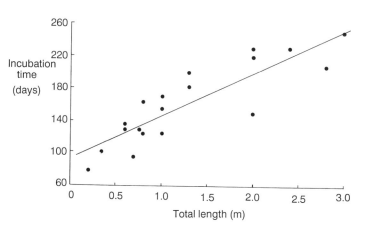

Figure 4.9
Incubation time and body size in varanids

low and fluctuating temperatures experienced by clutches in natural situations. Incubation times are longer under lower temperatures (26°C to 28°C) than those at higher temperatures (30°C to 31°C).

The breeding timing of different species varies considerably. In areas of northern Australia, some species breed in the wet season while others in the same region breed in the dry season. Asian forest-dwelling species (V. bengalensis, V. dumerilii, V. indicus, V. prasinus, V. rudicollis, V. salvadorii and V. timorensis) lay their eggs between August and March while Australian desert-dwelling species (V. acanthurus, V. brevicauda, V. eremius, V. gouldii, V. giganteus, V. spenceri and V. tristis) lay their eggs between September and January. The onset of breeding, even in those species which breed in the same season, may differ by several months. Populations of the same species inhabiting different climatic regions may start breeding several months apart. More than one factor may determine the time of breeding. Females require an adequate food supply to be able to produce eggs and hatchlings need a good food supply to survive and grow rapidly. Soil conditions must be suitable to enable the female to dig a nest that provides a suitable microclimate for the eggs. There is also variation in time of breeding within widespread species such as V. salvator, V. gouldii and V. tristis which live in areas with large climatic differences.

HATCHING

The eggs of V. rosenbergi and V. varius incubate in termitaria over winter and hatch (Plate 9) in early spring, about eight months after they are laid. Hatchlings of other species emerge at various times of the year, when conditions are favourable. Those of V. acanthurus and V. brevicauda and probably those of other pygmy goannas, appear in early summer (January) after approximately two and a half months of incubation. Those of the medium-sized species V. tristis hatch in late summer (March) after three and a half to four months incubation. In eastern Indonesia, hatchlings of V. komodoensis emerge at the start of the dry season, while those of V. timorensis apparently emerge at the end of the dry season.

Hatchlings of different species vary greatly in size. *V. rosenbergi* hatchlings weigh about 18 grams and are approximately 170 millimetres long while those of *V. albigularis* weigh 21 grams and are over 240 millimetres long. Those of the pygmy species *V. gilleni* and *V. storri* are much smaller, weighing 3 grams and being 130 to 135 millimetres long. The hatchlings of large species are much larger. Those of *V. komodoensis* weigh between 80 and 127 grams and are about 300 millimetres long, while hatchlings of *V. salvadorii* weigh 55 grams but can be up to 490 millimetres long due to their very long tails. The relative lengths of the body and tail in hatchlings can vary greatly between species.

Newly hatched *V. rosenbergi* (Plate 9) do not immediately leave the termitarium. They spend several weeks within the nest (Plate 10), slowly excavating an escape tunnel that follows the path of entry made by the mother prior to egg-laying. Even when the escape tunnel reaches the surface of the mound the young do not emerge immediately, but await the arrival of sunny and warm weather. They then emerge to briefly bask on or near the mound before moving away to forage (Plate 11). They make frequent returns to the nest throughout the day, and remain there overnight for up to several months after hatching. Most *V. rosenbergi* hatchlings fall prey to predatory birds and snakes within a few weeks of emergence from the nest.

Whereas the young of *V. rosenbergi* and *V. niloticus* are able to dig their own way out of termitaria, the escape of *V. varius* hatchlings from termitaria is assisted by adults that dig into the egg chamber. The hatchlings of *V. varius* are very active and adept at climbing, using their prehensile-like tails. They spend much of their time in trees, like the juveniles of other large species such as *V. komodoensis* and *V. bengalensis*. They are less likely to be observed there than animals that are active on the ground. Varanids are cannibalistic and they and other predators readily capture and eat hatchlings of all varanid species.

Varanid hatchlings are generally much more colourful and strongly patterned than the adults, although some adults of some species are quite colourful, for example *V. glauerti* (Plate 13) and *V. salvator cumingi* (Plate 12). The hatchlings of *V. rosenbergi* (Plate 10) are black and blue with red and white on their throat and head. The hatchlings of *V. glauerti* (Plate 14), *V. storri* (Plate 15), *V. dumerilii* (Plate 16), *V. tristis* (Plate 17), *V. pilbarensis* (Plate 18) and *V. prasinus* (Plate 19) vary considerably but all are brightly coloured. The reasons for the striking colouration of the young are unknown, but it may serve either as camouflage or as a deterrent to predation, as many toxic species of animals use very bright colouration as a warning to predators.

GROWTH RATES

Few data are available on the growth rates of young *V. rosenbergi* due to the high mortality rate of hatchlings and the absence of captive animal data. However, a few young recaptured six months after hatching had increased from about 20 grams to 100 grams. The early growth rate data that are available for a few other species of goannas are almost totally limited to those from captive animals. Growth rates are probably influenced by the conditions under which most animals are kept, particularly by the amount of food provided. In most cases, the young captive goannas probably received more food than would have been available to wild animals.

The duration of growth seems to be influenced by the adult size of the species. In captivity, the young of some species rapidly increase their weight (by two to four times) and their total length (by one to three times) during their first three to five months of life. After one year some of the smaller species are approaching their maximum length (approximately twice that at hatching), whereas the length of the larger species, such as *V. varius* (five to six times hatchling length) and *V. komodoensis* (up to ten times hatchling length) continues to increase until they are well over five years old.

Nor do we know at what age *V. rosenbergi* starts to breed. The estimated age of sexual maturity for *V. komodoensis*, based on data from captive animals is approximately five years and that for *V. bengalensis* is three years for males and four years for females.

Table 4.2 Longest known periods in captivity of varanids

Species	Period
V. albigularis	16.7 years
V. bengalensis	5–6 years
V. exanthematicus	17+ years
V. flavescens	10 years 6 months
V. gouldii	approx. 7 years
V. griseus	9 years 6 months
V. komodoensis	16 years 7 months
V. niloticus	15 years
V. salvator	10 years 8 months

Note: Age of animals was unknown at beginning of captive period.

LONGEVITY

Few data are available on the longevity of varanid lizards, as no long-term field studies have been carried out to determine the population dynamics and survival rates of any species. The longest interval between the first and last capture dates of an individually marked *V. rosenbergi* on Kangaroo Island was just over six years. When first captured, it was a 1290 gram adult with a 420 millimetre snout–vent length. It did not increase in length during the following 74 months and its weight only increased to 1320 grams. No data are available on the maximum survival time of *V. rosenbergi* in captivity. The longest time between captures of a wild *V. komodoensis* is eight and a half years.

At least nine species of varanids are known to have lived for more than five years in captivity. Records of the longest periods during which species have been kept in captivity are shown in Table 4.2.

GENERAL BEHAVIOUR

HOME RANGE

The home range of an animal is the total area used for all of its activities. Little is known about the sizes of the home ranges or activity areas of goannas, although accounts of their activity levels generally state that most species are wide-ranging, active foragers.

The home ranges of 13 *V. rosenbergi* on Kangaroo Island were determined by recapturing animals over periods ranging from 5 to 74 months. The capture locations were plotted on a map, the outermost points were connected and the areas within were measured (Figure 5.1). The home range areas for these animals (Table 5.1) were highly variable, as has been found in the limited data on home ranges for other varanids (*V. giganteus, V. gouldii, V. tristis, V. varius, V. albigularis, V. komodoensis* and *V. griseus*). The reasons for this variability are unknown, but it may be due to behavioural differences between sexes, age, season when data were collected or/and length of data collection period, or to variations in hunting ability and availability of prey. The sex of most species of goannas, including *V. rosenbergi*, is very difficult to determine with certainty on the basis of external morphology. In the case of goannas on Kangaroo Island, it was not possible to relate sizes of the home range to their sex. The home ranges of some *V. rosenbergi* overlapped to a considerable extent (Figure 5.1). Individuals of *V. rosenbergi* and *V. giganteus* also use different areas in different seasons.

There is an empirically derived equation which predicts the size of the home range of a lizard on the basis of its body weight. The values determined for the home ranges of *V. rosenbergi* are greater than those predicted by this method, as occurs with many species of *Varanus*. The home ranges of males are often 10 to 30 times those that are predicted on the basis of their body weight. The maximum sizes of home ranges are often a result of high levels of movement by males (*V. albigularis*, *V. tristis*) searching for receptive females during the breeding season. However, the predicted home ranges are based mainly on data obtained from herbivorous and insectivorous species of lizards, which are thought to use smaller areas than carnivorous species. All large varanids are carnivorous.

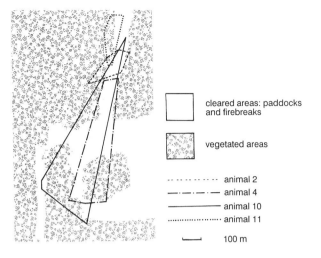

Figure 5.1 Home ranges of *V. rosenbergi*

ACTIVITY AREAS

An activity area is that portion of a home range used during a defined period such as a day or a season. On Kangaroo Island, *V. rosenbergi* were active away from their burrows on most suitable days throughout the year. The sizes of daily activity areas of goannas fitted with radio-transmitters were determined by plotting their locations several times each day. The activity areas (which were calculated in the same way as home ranges) varied seasonally, the largest being used during summer and the smallest during winter (Table 5.2). During the warmer months, the animals moved furthest, as goannas then reached their highest body temperatures and were active for the longest periods each day. In the warmer months, goannas were commonly found on low-lying, cleared areas. During winter, when these areas become very wet, the goannas moved onto elevated, well-drained and more densely vegetated sites.

The small arboreal species *V. caudolineatus* has a mean maximum distance travelled of only 34 metres per day. It mainly shelters in the hollows of dead trees and forages on the ground during hot periods. The even smaller species *V. brevicauda* is also very sedentary and may be inactive for long periods during harsh environmental conditions.

GENERAL BEHAVIOUR

Table 5.1 Home range sizes of *Varanus rosenbergi* on Kangaroo Island

Animal number	Weight at last capture (g)	Months from first to last capture	Home range (ha)
1	770	10	7.22
2	990	24	4.14
3	1060	18	36.95
4	1370	28	37.52
5	1795	25	1.71
6	735	5	5.57
7	1350	32	14.21
8	1320	74	16.19
9	950	41	6.18
10	1465	46	43.70
11	700	32	3.36
12	1425	55	36.92
13	1010	58	38.99
Mean	1149		19.44±4.58

Table 5.2 Seasonal variation in the mean size of daily activity areas (in hectares) of *V. rosenbergi* on Kangaroo Island

Month	Days of observation	No. of animals	Daily activity area (ha)
June	10	2	18 (0.16 0.20)
August	14	3	0.71 (0.19–1.11)
December	17	3	1.37 (0.67–2.43)
March	17	2	0.66 (0.62–0.70)

Note: Values in brackets are the range.

ACTIVITY PATTERNS

The seasonal activity patterns of different species of goannas can differ greatly. Those living in temperate or desert regions are often inactive for long periods during the winter or in drought conditions, and most tropical varanids also become inactive in the late dry season.

The daily activity pattern of goannas in temperate or tropical areas normally consists of a single (or unimodal) period of activity during the day in seasons when they are active, but there is some variability between species and individuals. The lizards usually emerge from their overnight shelter in the morning and remain outside until the late afternoon or evening and then retreat back into shelter. On Kangaroo Island the activity period is longer during the summer than in the cooler months. Desert-dwelling species, such as *V. gouldii* and *V. giganteus*, generally have two periods of activity (bimodal) during the day in the hotter months. Desert forms emerge earlier in the day than they do in cooler seasons and retreat to their burrow, or into some other shelter site, in the middle of the day when it is hottest. Some goannas emerge again for a second period of activity during the afternoon. During the cooler months they may adopt a unimodal activity pattern or remain inactive in their burrows. They may not emerge from their burrows at all during winter, perhaps because of a lack of available prey, or because low soil or burrow temperatures do not provide the stimulus for them to emerge. Some species (*V. glebopalma, V. tristis*) have been observed being active after dusk on hot evenings.

In some species of goannas (*V. tristis, V. albigularis* and *V. bengalensis*) and other lizards, males have larger territories and are more active within them than are females. As most goannas are encountered by chance when they are active, those individuals which are most active will be those most frequently encountered. Sex ratios in museum collections of many species of varanids are strongly biased in favour of males (*V. eremius, V. glauerti, V. glebopalma, V. semiremex, V. caudolineatus, V. storri, V. tristis, V. scalaris, V. gouldii, V. panoptes, V. mertensi, V. mitchelli, V. indicus, V. varius, V. komodoensis, V. salvator* and *V. rosenbergi*). These are thought to be due to differences between the sexes in activity levels. A sample of 45 dead *V. rosenbergi* from Kangaroo Island consisted of 37 males and 8 females, a ratio of 4.6:1, which was probably due to differences in the levels of activity between the sexes. However, because of the difficulty in determining the sex of live goannas of this species, it is not definitely known whether these behavioural differences occur between male and female *V. rosenbergi*. Another species, *V. acanthurus*, is generally captured at its shelter sites. There is no male bias in museum collections of this species, since there do not appear to be any differences in activity levels between the sexes to influence the sex ratio of captured animals.

FORAGING STRATEGIES

Varanids are intensive foragers that spend much of their time and energy searching for prey and investigating their environment. This requires acute sensory capabilities and a highly developed mental capacity to enable them to locate infrequent and patchily distributed prey. Different strategies are needed for different habitats and prey types. Foragers like varanids need a good memory and must be able to associate different stimuli (such as scent, remembered traits and search images) with each type of prey in order to form appropriate foraging strategies for them.

Varanids sometimes use complex strategies to select particular prey items to maximise the amount of energy they can derive from their food. For example, *V. niloticus* have been observed to cooperate when foraging. After locating a crocodile nest, one member of a pair appears to lure the female crocodile, which is guarding the nest, away from it. This allows the other to open the nest chamber to expose the eggs which it then feeds on. The decoy varanid then returns to the crocodile nest and also feeds on the eggs.

The memories of varanids are apparently very good. Some *V. komodoensis* live in the core areas of their home ranges for at least 15 years. During that time they learn and remember reliable watering places, as well as the locations of sleeping sites and game trails of the deer and pigs on which they prey.

A highly interesting observation was made recently during feeding experiments on captive *V. albigularis*. Varanids can count! The lizards were conditioned to feed on groups of four snails which were contained in separate compartments with movable partitions. These were opened one at a time, allowing the varanids to eat all snails in each group. Once the lizards were conditioned, one snail was removed from some groups of snails. Not being able to find the 'missing' snail in a group always caused the lizards considerable distress, and they searched extensively for it even when they could see the next group of snails. They were able to determine that one snail was missing from groups of up to six snails. With larger groups of snails, the lizards seemed to classify them simply as 'lots', ate them all, then moved on to the next cluster. Therefore, varanids can count up to six, but after that it apparently does not matter, or they simply cannot cope with the numbers. So it would seem goannas can count better than a lot of small children.

BURROW USE

The entrance to a *V. rosenbergi* burrow is frequently situated beneath a flat rock, a small shrub or a fallen log (Figure 5.2a). There is usually a mound of earth at the burrow entrance and the animal often lies upon this to bask for some time after emerging from the burrow. The burrows of *V. gouldii* commonly have a small shaft that runs from the terminal chamber of the burrow up towards the soil surface. If the goanna is within the terminal chamber and a predator begins to enter the burrow, the goanna can escape through this 'pop-hole' (Figure 5.2b).

In most cases, goannas dig their own burrows and normally live alone in them. However, some *V. rosenbergi* occasionally share a burrow with a non-breeding animal of the same species, and at other times a burrow will be occupied by different goannas on different nights. Individual *V. rosenbergi* usually frequent more than one burrow, but there is no detailed information on the frequency of use of a burrow or the numbers of individuals that use a particular burrow.

The warrens of introduced rabbits (*Oryctolagus cuniculus*) occur widely within the ranges of *V. gouldii*, *V. rosenbergi*, *V. panoptes* and *V. varius*. These goannas frequently use rabbit warrens for shelter instead of constructing their own burrows.

Burrows not only play an important role in protecting a goanna from predators, but they are also very important in providing protection against the weather (see Chapter 6, Thermal Biology).

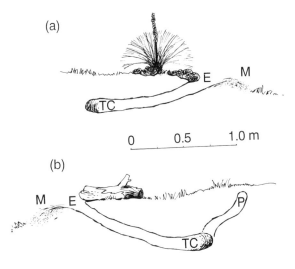

Figure 5.2
Burrows of
a. *V. rosenbergi*
b. *V. gouldii*.
Legend:
mound (M)
entrance (E)
terminal chamber (TC)
'pop-hole' (P)

GENERAL BEHAVIOUR

MOVEMENT AND BODY POSTURES

When walking, goannas carry their bodies high off the ground and only a small part of the tail near the tip touches the ground surface. This gives rise to the typical goanna tracks seen in sandy or dusty soils (Figure 5.3). When running, the tail is lifted clear of the ground (Figure 5.5, page 46).

Figure 5.3
Tracks of a varanid

In its usual, non-aggressive stationary posture a varanid has most of its body resting on the ground while the head and shoulders are slightly raised (Figure 5.5). When the animal adopts an aggressive stance, it remains on all fours, with the gular pouch and abdomen inflated (Figure 5.5b and c, Plate 20). The goanna hisses loudly and can produce powerful side swipes of the tail (Figure 5.5b and c). In several species, including *V. giganteus*, these side swipes are said to be capable of breaking the legs of dogs. The most aggressive posture in varanids is where the animal stands upright, inflates the gular pouch and makes loud hissing sounds from the nostrils (Figure 5.5d). The animal may sway slightly. A similar posture is used in ritual combat by some species (Plate 7). Aggressive postures are directed not only towards members of the same species but also at potential predators, such as humans, and sometimes at venomous snakes. Threat displays such as inflation of the gular pouch are also shown by hatchlings of some species.

Figure 5.4
Running
V. komodoensis

GENERAL BEHAVIOUR

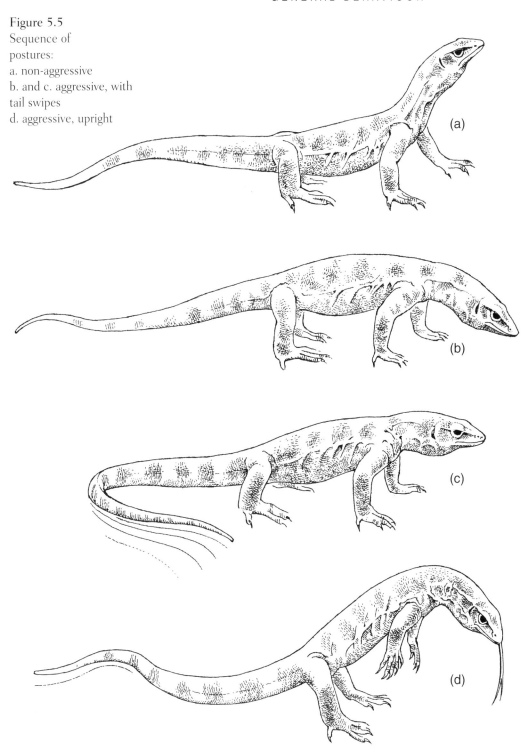

Figure 5.5
Sequence of postures:
a. non-aggressive
b. and c. aggressive, with tail swipes
d. aggressive, upright

A common posture in some large varanids is a bipedal stance in which the animal raises itself vertically and supports itself on its hindlegs and tail (Figure 5.6 and Plate 7). The animal then has the clear and elevated view of its surroundings, and this posture may have given rise to the 'monitor' name.

The behaviour patterns associated with hunting and breeding are discussed in Feeding (Chapter 3) and Breeding (Chapter 4).

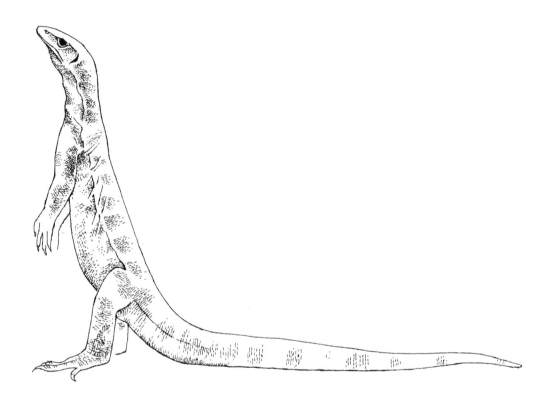

Figure 5.6
Upright stance of
V. panoptes

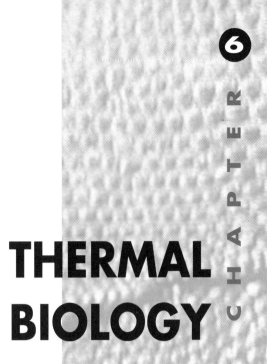

THERMAL BIOLOGY

REGULATION OF BODY TEMPERATURE

All varanids are ectotherms. That is, they heat their bodies by basking and absorbing much of their energy directly from the rays of the sun or by conduction from warm surfaces. Goannas of different species maintain their bodies within a temperature range of 30°C to 40°C when they are active (Figure 6.1), although semi-aquatic species mainly operate at body temperatures near the lower end of that range. Most species regulate their body temperatures to within two or three degrees of 36°C while active. The body temperatures of animals resting in their burrows are lower than their activity temperatures and can drop to below 10°C in the winter near the northern and southern limits of their distribution or in desert areas.

BURROWS

Burrows are often used by goannas for protection against overheating on hot days. However, burrows and other forms of shelter, such as holes in trees, hollow logs and brushpiles, also allow body heat to be conserved overnight. The animals can therefore emerge from their shelters in the morning at a higher temperature than their

THE BIOLOGY OF VARANID LIZARDS

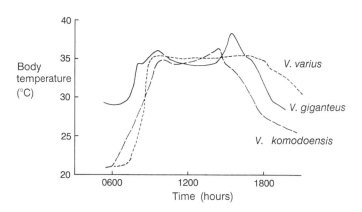

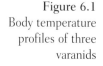

Figure 6.1
Body temperature profiles of three varanids

surroundings, and be more active and alert than would otherwise be the case.

Burrows of *V. rosenbergi* on Kangaroo Island are relatively shallow, slightly curving, and have an enlarged terminal chamber at the deepest point (Figure 5.2a, page 44). Burrows are usually located on flat, open areas during spring and summer. Throughout the year, burrows can also be found on sandy ridges, but during the winter only burrows at these elevated sites are inhabited. The average length of burrows on Kangaroo Island is 1050 millimetres and does not vary seasonally, although the depth of the burrow does. Those used during the summer have an average depth below the soil surface of 137 millimetres, while those used during winter are significantly deeper with an average depth of 180 millimetres. The burrows excavated by *V. gouldii* during summer in semi-arid habitats are over 400 millimetres below the soil surface. Below soil depths of 150 millimetres, the temperature within the burrow fluctuates very little.

Burrows are used as shelter by many species of goannas (*V. eremius, V. gouldii, V. griseus, V. salvator, V. panoptes, V. bengalensis* and *V. komodoensis*). They vary in shape and are generally relatively short and shallow. Often they are dug by other animals and taken over by goannas. Some species also shelter in hollow trees (*V. varius, V. scalaris, V. caudolineatus, V. tristis, V. komodoensis*), in abandoned termite mounds (*V. bengalensis*) or beneath the bark of dead trees (*V. caudolineatus, V. gilleni*).

The temperatures in the burrows of *V. rosenbergi* vary only slightly during the day, reaching a maximum between 4 pm and 10.30 pm, and a minimum temperature between 6 am and 9 am. Burrow temperatures are more stable than air temperatures immediately outside the burrows (Figure 6.2). In summer, burrows provide cool refuges during the heat of the day, while in winter they remain warmer than

Figure 6.2
Differences between environmental and body temperatures in *V. rosenbergi*

the cold night air. At a given time of year, the burrow temperature is approximately the same as the daily mean air temperature outside the burrow. The relative humidity in the burrows remains at or near saturation level even during summer, thereby assisting the goannas to reduce their evaporative water losses.

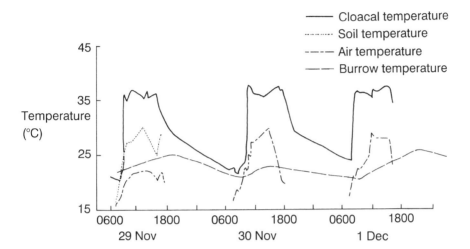

EMERGENCE TIME

V. rosenbergi on Kangaroo Island emerge from their burrows earlier during summer than in winter. Mean time of emergence during summer is 8.50 am (ranging from 7.35 to 10.10 am) while in winter the mean time of emergence is 9.37 am (ranging from 7.45 to 10.45 am). On about 10 per cent of winter days goannas on Kangaroo Island do not emerge from their burrows at all. On most winter days they move to the mouth of the burrow in the morning and lay there in apparent anticipation of becoming fully active. The mean body temperature at the time of emergence in summer is 21.8°C (ranging from 15.0°C to 24.6°C) while the mean emergence temperature in winter is 12.5°C (ranging from 9.5° to 15.0°C). Temperatures within the burrow are similar to the body temperature of the lizard at the time of emergence. The thermal chambers of desert-dwelling populations of *V. gouldii* are often closer to the surface than the rest of the tunnel, which may enable them to monitor the surface temperature.

The precise stimulus for goannas to emerge from their burrows is unknown. It appears that the difference between the temperature inside the burrow and the increasing soil temperature near the burrow entrance may be an important factor. The time they emerge may also vary according to the location or orientation of the burrow. Those animals using burrows near the tops of ridges, or with entrances facing east, generally emerge earlier than goannas using burrows with other locations or orientations. The conditions and times of emergence of other species of varanids are also variable (*V. griseus, V. gouldii, V. giganteus, V. komodoensis* and *V. varius*).

BASKING

Basking allows the body to gain heat using the radiant energy of the sun. It generally occurs in the morning, after the lizards have emerged from their overnight shelter, but may also occur at intervals throughout the day. Goannas align their bodies in such a way as to maximise heat gain, or elevate themselves above the cold ground by climbing onto logs or rocks to present a larger surface area to the rays of the sun. The heating rates of some species, including *V. rosenbergi* (Table 6.1), vary seasonally. On sunny winter days, the periods of basking are longer and the rates of heating are slower than in the warmer months.

It is a common practice of many lizards to protrude only their head from the burrow before emerging completely. It has been suggested that this facilitates an increase in the temperature of the head compared with that of the body, which may increase the level of mental activity of the lizards before they emerge fully. No heating of the body occurs at this time, no doubt because the body has a much greater mass and thermal inertia than the head and there is little transfer of heat from the head to the body.

When a varanid has attained a sufficiently high body temperature by basking, it can move away from the vicinity of its burrow. While active, lizards travel through patches of sunlight and shade. The density of the vegetation on much of Kangaroo Island provides abundant shade for *V. rosenbergi*. Because the goannas are unable to produce sufficient internal body heat, and have little insulation to help retain it, their body temperatures fluctuate slightly during activity periods. Nevertheless, it is advantageous for ectotherms to tolerate minor fluctuations in body temperature. If they were unable to do so, they would waste time and effort in frequent shuttling between sunny and shady areas to regulate their body temperature more finely. This is especially so in smaller species because the temperatures of these animals vary more widely due to their high body surface to mass ratio.

Table 6.1 Seasonal differences in heating rates and times of basking of *Varanus rosenbergi* on Kangaroo Island

Month	Heating rate (°C/minute)	Basking time (minutes)	Air temperature (°C)
May–June	0.15 (0.04–0.20)	102.5 (85–130)	18.2 (16.3–20.4)
August	0.10 (0.02–0.18)	130.3 (48–265)	15.6 (12.6–19.7)
November–December	0.27 (0.11–0.44)	61.4 (35–95)	22.1 (18.0–26.1)
March	0.28 (0.17–0.49)	60.0 (15–110)	29.5 (23.3–33.7)

ACTIVITY TEMPERATURES

Two methods are used for studying the temperatures of free-ranging lizards. Radio-telemetry (the use of temperature-sensing transmitters) enables a number of temperatures from one animal to be obtained every few minutes while the animal is active in its normal environment. This allows the temperature regulation of an individual to be examined. The other method involves the use of thermometers and requires the frequent recapture of an individual. This may alter its behaviour pattern while it is being observed between captures. To overcome this, a number of animals are captured only once each. Consequently, data obtained using thermometers are of less value than transmitter data, as they provide only an average of single values obtained at the time of capture from a number of different individuals.

Under favourable conditions, active *V. rosenbergi* on Kangaroo Island maintain a mean body temperature of 35.6°C, ranging from 32.1° to 38.0°C (Figure 6.2, page 51). These temperatures were determined by using radio transmitters on three goannas, releasing them into the wild, and recording their body temperatures while they were active over several days. The mean body temperature of active animals measured with a thermometer shortly after capture during summer is 35.1°C (ranging from 32.4° to 37.6°C).

The semi-aquatic species *V. salvator* in Malaysia appears to prefer an activity temperature of 30°C to 32°C, but can function well at any body temperature between 29°C and 36°C. This species frequently spends time in the ocean and may do so to reduce its body temperature during hot parts of the day. The mean activity temperatures of three free-living semi-aquatic varanids (*V. niloticus*, *V. salvator* and *V. mertensi*) are several degrees lower than those of the more terrestrial species. These animals are frequently submerged in water that is much cooler than the air and the animals would have to spend excessive periods basking in order to maintain the higher body temperatures commonly found in terrestrial species.

The mean activity temperatures of gila monsters (*Heloderma suspectum*) and beaded lizards (*Heloderma horridus*), which are in a family closely related to the varanids, are 28.5°C (range 22.5 to 36.0°C) and 29.3°C (range 17.4 to 36.8°C) but these species are much less active than varanids as they are active on only 10 per cent of summer days and only move over short distances when foraging.

Studies of temperature regulation of reptiles in the laboratory are generally conducted in artificial temperature gradients. These usually consist of a narrow cage up to several metres long with a heat source at one end. This produces a gradation in air and surface temperatures along the length of the cage. The animal is thus able to select a temperature at any position along the gradient and to regulate its body temperature at its desired level. The thermal environment can be changed to allow particular aspects of the animal's behavioural and physiological reactions to be observed.

The mean body temperature selected by active *V. rosenbergi* in temperature gradients is 35.4°C, which is very similar to the temperatures selected by this species in the wild. The mean body temperatures of individuals have been observed to vary by several degrees on different days or during different seasons in a number of varanid species.

Lizards in the wild often become active at lower body temperatures than those selected by captive animals in thermal gradients. Their body temperatures are influenced by the prevailing ambient temperature and degree of solar radiation (insolation). During the colder months, lower air temperatures and less intense solar radiation may prevent them from reaching their normal activity temperatures (Table 6.2). In summer, goannas avoid becoming overheated by seeking shade, as excessive body temperatures (over about 43°C) will cause death. Free-ranging goannas minimise the range of environmental conditions they encounter by choosing the appropriate time of day to be active.

The activity temperatures of *V. rosenbergi* are similar to those of most other terrestrial or arboreal varanids. The temperatures of active, free-ranging goannas of eight other species (*V. giganteus*, *V. varius*, *V. griseus*, *V. gouldii*, *V. komodoensis*, *V. salvator*, *V. scalaris* and *V. tristis*) have been determined using radio-telemetry (Table 6.3). Values determined for other terrestrial or arboreal species using single values obtained with a thermometer are shown in Table 6.4. The body temperatures of active varanids are similar in species that vary greatly in body size.

Table 6.2 Seasonal differences in activity temperatures of *Varanus rosenbergi* on Kangaroo Island

Month	Number of values*	Mean body temperature (°C)	Mean air temperature (°C)
May–June	34	29.3 (18.4–36.8)	15.8 (14.2–18.3)
August	28	24.5 (17.1–37.5)	14.5 (12.4–16.3)
November–December	179	33.4 (21.3–39.2)	22.3 (16.2–33.0)
March	36	32.0 (27.5–36.3)	29.5 (18.5–37.7)

*sample size

Table 6.3 Activity temperatures of eight species of free-living terrestrial and arboreal varanids, as determined by telemetry

Species	Mean temperature (°C)	Temperature range (°C)
V. giganteus	35.8	26.8–39.4
V. gouldii	35.5	27.2–38.1
V. griseus	36.8	27.0–41.6
V. komodoensis	35.1	27.6–41.3
V. varius	35.5	32.8–36.4
V. salvator	30.4	27.7–32.8
V. scalaris	38.9 (wet season)	33–40
	35.6 (dry season)	33–40
V. tristis	37.5	30.5–40.2

Table 6.4 Activity temperatures of twelve species of terrestrial and arboreal, and three species of semi-aquatic, varanids as determined from single values taken by thermometer

Species	Free-living (°C)	Captive (°C)	Total length (m)
TERRESTRIAL OR ARBOREAL			
V. bengalensis	32–37	—	2.0
V. caudolineatus	37.8	—	0.25
V. eremius	37.5	35.9	0.45
V. exanthematicus	36.4	36.5	2.0
V. giganteus	36.1	—	2.4
V. gilleni	37.4	37.1	0.7
V. gouldii	37.0	35.3	1.0
V. griseus	38.5	36.4	1.3
V. komodoensis	36–40	36.3	2.8
V. rosenbergi	35.1	35.2	1.5
V. tristis	34.8	35.4	0.8
V. varius	34.7	33.5	2.0
SEMI-AQUATIC			
V. mertensi	32.7	32.5	1.3
V. niloticus	32.7	34.8	2.0
V. salvator	27–32	35.6	2.5

HEAD–BODY TEMPERATURE DIFFERENCES

Some researchers believe that lizards regulate their head temperatures more carefully than their body temperatures, in order to prevent overheating of the brain. Temperature gradients occur within the bodies of reptiles and they are maintained by adjustments to blood circulation and respiration. Thermal inertia of tissues and differences in the surface area to volume ratios of different parts of the body are also involved.

Head and body temperatures in several V. rosenbergi and V. gouldii in thermal gradients were measured simultaneously. The temperature inside the skull of the goannas was generally slightly lower than the cloacal temperature before the animals began basking. After the lizards emerged from artificial burrows into the gradient and started to bask beneath heat lamps, their head temperatures increased more rapidly than their body temperatures. After 20 to 40 minutes of basking, body temperature rose above cranial temperature and remained above it for the remainder of the animals' activity periods. When heat was no longer applied, the heads and bodies of the lizards cooled at similar rates (Figure 6.3). When measured over a period of weeks, the average internal head and body temperatures during activity were similar (Figure 6.4), as were the maximum temperatures.

If only the body of a goanna is heated, both head and body temperatures rise, whereas if only the head is heated its temperature rises while the body temperature does not. This may explain why goannas protrude their heads from their burrows for a short time before they emerge completely.

A study using transmitters on free-ranging V. varius showed that brain temperatures were higher than body temperatures by as much as 6°C before emergence, or during the initial period of basking. However, once the animal became active, brain and body temperatures were similar.

The differences in the mass to surface area ratios between the heads and bodies of varanids and the blood flow between them are responsible for the differences in the rates of heating of the head and body. Evaporative cooling of the moist surfaces of the mouth and the eyes also affects the temperature of the head.

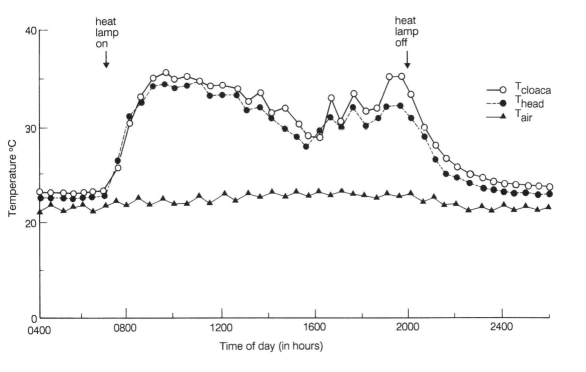

Figure 6.3
Head and body temperatures of V. rosenbergi while basking and active during one day

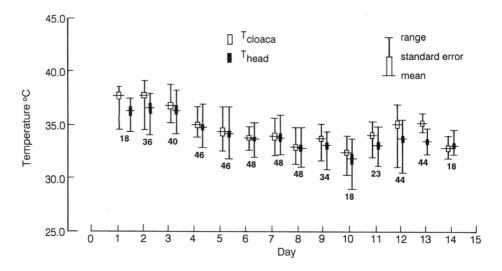

Figure 6.4 Average head and body temperature differences of active *V. rosenbergi* over two weeks in a thermal gradient

NASAL TUBES

The anterior nasal tubes have a large surface area and this is important in heat exchange and water conservation. Inhaled air is warmed and humidified as it passes along the nasal tubes. As exhaled air passes back along the nasal passageways, which are now cool, heat is transferred from the air back to the tissues. Water vapour is condensed onto the surface of the nasal tubes before it leaves them. Thus water loss is less than it would be if expired air remained at deep body temperature. In some species, including *V. rosenbergi*, the nostril is near the tip of the snout. This means the nasal tubes are as long as possible, which aids in warming the inhaled air. In some other species the nares are immediately in front of the eyes, and the nasal tubes are short.

GULAR FLUTTERING

Gular fluttering, which is often erroneously referred to as panting, is used by goannas and by many other types of lizards to cool themselves when they become overheated. The mouth is held open and the gular region (Figure 1.1, page 6) of the throat is fluttered rapidly. This has the effect of increasing the evaporation rate from the mouth and throat area and therefore reducing temperature. However, it is not often employed as a means of cooling because water losses may increase by four to six times when it occurs (see Chapter 8, Water Use, and Figure 8.1, page 67). Nevertheless, it is a means of reducing dangerously high head temperatures in an emergency. Lizards normally avoid these problems by retreating from stressful temperature conditions to areas of shade, or into burrows.

Plate 1
Rosenberg's monitor (*V. rosenbergi*) from Kangaroo Island, South Australia. They are the largest terrestrial predator on the island and are very abundant there.
(Brian Green)

Plate 2
The Komodo dragon (*V. komodoensis*), from several islands in eastern Indonesia, is the largest living species of lizard.
(Dennis King)

Plate
This speci
which lives
northern Austra
is often referred
as V. 'pellewens
which is not
official nam
It will probably
described a
new species
the near futu
(John Womb

Plate
The jaw bones a
muscles
Komodo drago
(V. komodoen
are power
enough to ena
the dragons to t
apart and eat la
prey, such as pi
deer and buffa
(Ron Johnsto

Plate 5
The semi-aquatic species *V. mertensi* hunts mainly in rivers and billabongs for fish and crustaceans, though it will also forage on land.
(Graeme Chapman)

Plate 6
The perentie (*V. giganteus*) live in central Australian deserts. The diet of adults consists mostly of large prey items.
(Hans-Georg Horn)

Plate
The extend
tongue of t
V. panoptes fr
northern Austra
can sample a v
large area for sce
particles. T
assists the
in locating fo
or mat
(Harry Butl

Plate 8
Nest site of
V. rosenbergi in a
termite mound on
Kangaroo Island,
South Australia
(Brian Green)

Plate 9
The young of
V. rosenbergi hatch
in the spring,
often in termitaria
where the eggs
incubate over win-
ter at a relatively
high and constant
temperature.
(Graeme Chapman)

Plate
Hatchlings *V. rosenbergi* often continue to shelter in the termitaria for weeks after hatching (Mike McKelvey)

Plate 11
The reason for the light coloration of hatchlings of *V. rosenbergi* is unknown, but may provide camouflage as they shelter on the forest floor.
(Brian Green)

Plate 12
Large numbers of the colourful skins of *V. salvator cumingi* are exported each year from the islands of Samar, Leyte, Bohol and Mindanao in the Philippines.
(Maren Gaulke)

Plate
Adult V. glauer
particular
those from th
Kimberley regio
of Weste
Australia, are mo
colourful tha
most other speci
of goanna
(Robert Jenkin

Plate
V. glauerti live o
rocky outcrops
Western Austral
and are arbore
in the Northe
Territo
They have tw
colour form
(Ber
Eidenmülle

Plate 15
V. storri occur in two distinct areas of inland northern Australia. They reproduce readily in captivity. (Bernd Eidenmüller)

Plate 16
V. dumerilii are secretive animals from southeast Asia. They seem to specialise in feeding on crabs. (Hans-Georg Horn)

Plate 1
V. tristis occur over most of northern and central Australia and have several different colour forms. They sometimes mate in trees. (Bern Eidenmüller)

Plate 1
V. pilbarensis have a very restricted distribution. They are only found on rocky slopes in the Pilbara region of northwestern Australia. (Bern Eidenmüller)

Plate 19
... prasinus is a slender arboreal species complex found on Cape York in northeastern Australia, in Papua New Guinea and on small islands of that region.
(Bernd Eidenmüller)

Plate 2 An aggressive posture is taken by *V. gouldii* in sand dunes near Lake Eyre, South Australia. Many species of goannas adopt similar postures when threatened. (Dennis King)

Plate 2 Some large species of goannas, including *V. mertensi*, engage in a bipedal form of ritual combat during the mating season. (Richard Braithwaite)

When *V. rosenbergi* are placed in temperature-controlled cabinets with rising temperatures, they begin gular fluttering when the body temperature rises above head temperature (about 38°C). This is similar to the temperatures at which *V. gouldii*, *V. griseus* and *V. komodoensis* begin gular fluttering. Gular fluttering has also been observed in *V. salvator* on hot days, but the body temperature at which it occurs is not known. Gular fluttering is not always continuous, but usually occurs at a rate of 60 to 90 per minute and reaches up to 120 to 130 per minute. Its onset is occasionally preceded by a period of struggling by the lizard. Head temperatures either stabilise or drop slightly when gular fluttering begins, although body temperatures continue to rise. Head temperatures rise during the brief periods when fluttering stops. Stable head temperatures can be maintained at as much as 2.8°C below body temperature for periods of up 75 minutes by means of gular fluttering and blood shunting.

Although head temperatures are maintained below ambient air temperature during gular fluttering, body temperatures remain the same as ambient. Since the depth of respiratory movements does not change during gular fluttering, it indicates that this activity is not respiratory, but serves to control the temperature of the brain.

REFLECTIVITY OF THE SKIN

Skin colour is important in thermoregulation as well as in breeding activity and camouflage. Some lizards can change their skin colour under different environmental conditions to increase or decrease the rates at which they take up or lose heat. Varanids, however, are unable to change colour or solar absorptance. Differences in skin colouration and reflectivity between species are related to the habitats they use. *V. rosenbergi* inhabits densely vegetated areas and is a darkly coloured (melanistic) species.

Nearly all of the light energy which reaches the Earth's surface is in the visible or infra-red parts of the spectrum. Some of it is absorbed by lizards when they bask. The amount of light reflected by skin from *V. rosenbergi* (from Kangaroo Island) and the light-coloured *V. gouldii* (from arid and semi-arid regions) has been compared. The greatest differences in reflectivity found in goannas from the three habitats are in the near infra-red portion of the spectrum (Figure 6.5). Skin from *V. rosenbergi* reflects less light than the others in all parts of the spectrum. Other varanid species from cool climates or with a semi-aquatic lifestyle, also show low levels of solar reflection from the skin.

Bare sandy soils reflect the most solar energy. Heavier, moist soils and those covered with heavy plant growth reflect the least solar energy. As there is usually abundant sunlight in deserts, temperatures

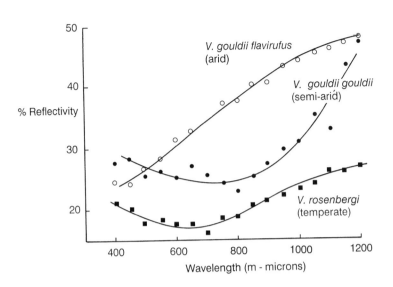

Figure 6.5 Light reflection from the skins of V. rosenbergi and V. gouldii

there are high. Lizards that live in such environments can easily become overheated, but they can remain active for longer periods if their skin reflects large amounts of this energy. The temperature on Kangaroo Island is generally lower than those in the arid and semi-arid regions of Australia, and the vegetation is also more dense than in the deserts. During much of the year there is not enough sunny weather to allow goannas to easily reach their activity temperatures when they bask. The dark skins of goannas living on Kangaroo Island absorb more solar energy, which means the animals can heat up more rapidly and be active longer in the colder, wetter months.

RESPIRATION

CHAPTER 7

Living cells require a constant supply of oxygen to enable them to respire and maintain normal cellular functions. Oxygen must be transported to the cells and the carbon dioxide produced in respiration must be carried away and expelled from the body. This process of gaseous exchange is carried out by the lungs and circulatory systems of animals.

In varanids, these systems differ in a number of ways from those of other reptiles. Reptiles, unlike mammals, do not possess a diaphragm (a sheet of muscle which separates the organs of the chest from the organs of the abdomen). In most reptiles the heart and lungs are not enclosed in a separate chest cavity but are located in the abdominal cavity with other organs (Figure 3.6, page 24). However, varanids and their relatives, *Lanthanotus* and *Heloderma*, have a septum (a dividing membrane) which separates the abdominal and chest (pleural) cavities. Except for crocodiles, reptiles do not seem to have specialised breathing muscles. They breathe in by drawing the ribs upwards and forwards by means of the intercostal muscles (situated between the ribs), increasing the size of the pleural cavity. When those muscles relax, the weight of the body and the elasticity of the muscles and other tissues forces air out of the lungs.

LUNGS

Reptile lungs have a volume approximately six times larger than the lung of similar-sized mammals, and are also larger than the combined lung–air sac systems of birds. However, the lungs of goannas are less elaborate than those of birds and mammals. Varanid lungs are larger than those of most other reptiles, thus enabling them to extract more oxygen from the air they breathe.

Reptilian lungs have to function over a wide range of oxygen consumption rates, due to the great differences in the animals' activity levels. Even so, these rates are much lower than those of birds and mammals which generally have much higher energy requirements.

Each lung has a cartilage-reinforced intra-pulmonary duct (the bronchus) which connects with a number of small chambers. This is called a multicameral lung (Figure 7.1). It provides an enlarged surface for gas exchange. This in turn provides an advantageous starting point for the evolution of a high-performance lung containing a large surface area which can be ventilated at a low energy cost. Exchange of oxygen and carbon dioxide between the inspired air and the air in the lungs is thus greatly enhanced.

Varanid lungs are attached directly to the body wall where it underlies the rib cage. This prevents them from collapsing completely because of over-stretching during normal breathing. They have a strong compliance with body movements and thus can be efficiently ventilated by means of costal breathing alone. This results from the normal movement of the ribs altering the volume, and thus the pressure, in the lungs. This and their multicameral structure produces a more efficient system of air circulation than that of other lizards. The lungs of most lizards lie free in the body cavity and must be actively deflated, thereby increasing the energetic costs of breathing. Varanids can extract oxygen from inspired air at rates that are similar to those of mammals at rest. Varanids also have a high aerobic capacity (capacity for aerobically supported activity which relies on oxygen) and breathing requires little extra expenditure of energy.

When active, most reptiles rely heavily on anaerobic respiration, which does not require oxygen. Instead, energy is released by the partial breakdown of stored carbohydrates to form lactic acid. Unlike aerobic respiration, anaerobic respiration has the advantage of being equally efficient over a wide range of body temperatures. Anaerobic respiration is therefore useful for reptiles which must be active before they have warmed up to their normal activity temperatures. However, it has the disadvantages of releasing only a small part of the energy available in the carbohydrate energy source, and of generating large quantities of lactic acid, which accumulate in the muscles and prevent sustained activity. Lactic acid is slowly removed from muscles by the blood and is later broken down aerobically. Active varanids, however, mainly rely on aerobic respiration and therefore do not generally build up high lactic acid levels in their tissues. They are therefore able to sustain high rates of activity over long periods of time.

The lungs of varanids receive a regular supply of blood, unlike those of other lizards, where blood is shunted away from the lungs during long periods of breath-holding. The breathing rate of varanids is generally regular and low, but it can be variable. Varanids can voluntarily hold their breath for long periods. For example, *V. exanthematicus* in the laboratory have been observed to hold their breath for up to 8 minutes. Even longer periods of breath-holding have been recorded for semi-aquatic varanids. At temperatures between 14°C and 20°C some small varanids can hold their breath for in excess of 30 minutes.

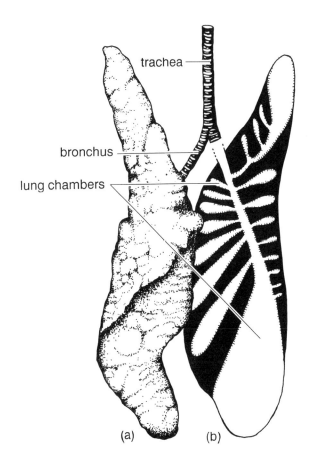

Figure 7.1
Lung morphology of a varanid:
a. entire
b. longitudinal section
(after Becker et al. 1989)

HEART

The heart in varanids is located further back in the body than it is in most lizards. It lies just behind the posterior part of the sternum (breastbone), embedded between the lobes of the liver. The anterior region of the heart is attached to a thick pericardium (a membranous sac which encloses the heart).

The varanid heart is not a typical lizard heart. The arrangement of the chambers and valves results in only a small amount of mixing taking place between the two bloodstreams (pulmonary and systemic). About 30 per cent of the blood intended for the lungs goes instead to the body and approximately 10 per cent of the oxygenated blood is returned to the lungs. This is a much lower level of mixing than found in most lizards. However, the structure of the heart appears to make some mixing of the bloodstreams unavoidable. The arterial (oxygenated) blood in varanids carries higher levels of oxygen than it does in most other lizards.

The relatively long necks in varanids may impose additional circulatory demands similar to those of snakes. Both types of animal require a supply of blood to the body at high pressure. However, the blood supply to the lungs is at a much lower pressure. These pressure differences are similar to those found in the circulatory systems of mammals. They are achieved by means of a muscular ridge in the ventricle which presses tightly against the heart wall at the appropriate phase of the heartbeat. This effectively creates a four-chambered heart and separates the two systems.

BLOOD PHYSIOLOGY

As the body temperatures of reptiles rise during activity, species that rely heavily on anaerobic respiration release large amounts of lactic acid into their blood. Increased acid levels raise the acidity of the blood unless an efficient buffering system is present to prevent it. This system must function effectively over a range of body temperatures. A high acidity of the blood reduces its oxygen-carrying capacity, which in turn increases the reliance of a lizard on anaerobic respiration. This in turn leads to the formation of yet more lactic acid which further decreases the oxygen-transporting ability of the blood.

The high level of aerobic respiration of varanids is partly due to the different way in which the buffers in their blood function. Varanids closely regulate the acidity of their blood as their body temperature increases and they become more active. The lungs of varanids are highly efficient in ridding the body of excessive carbon dioxide, which helps maintain the function of buffers to prevent an increase in blood acidity. Their lactic acid production is also maintained at low levels. Goannas are thus able to maintain their activity for long periods.

Varanids also have higher levels of myoglobin than other lizards. Myoglobin is a protein in the muscle which binds with oxygen transported there by the blood. These levels are similar to those in mammals, and improve the rate of oxygen transfer from the blood to the muscles.

The blood supply to the lungs also increases when body temperature rises. Thus varanids are very mammal-like in many features of their cardio-pulmonary physiology. This includes the need to maintain high levels of oxygen intake when they have high body temperatures but are at rest.

The respiratory properties of the blood of the semi-aquatic species of varanids show some adaptations to extended submergence. The buffering capacity of their blood allows them to be more tolerant of lactic acid accumulation than other species of varanids. High lactic acid levels will initiate breathing and thus restrict the length of diving periods. Remaining submerged for as long as one hour has been observed in some semi-aquatic species, as a result of their higher tolerance to lactic acid. During diving, the body temperatures of semi-aquatic species drop rapidly. To overcome this, the blood has properties which show distinct adaptations to diving and is able to release oxygen to the tissues at lower body temperatures than occurs in terrestrial species. When diving, the heart rate may drop by up to 85 per cent of its pre-dive level.

The advanced structure and performance of the heart and circulatory system, in conjunction with the complex structure of the lungs, allows varanids to respire efficiently for long periods without them becoming exhausted. Thus they are able to adopt an active, wide-ranging foraging strategy that is suited to their feeding habits.

CHAPTER 8

WATER USE

Six species of goannas live near water, and they can be regarded as essentially semi-aquatic: *V. indicus, V. jobiensis, V. mertensi, V. niloticus, V. salvator* and *V. semiremex*. These species are usually found in freshwater habitats, but can extend into brackish waters, or even sea water. This is especially true of those monitors which inhabit mangroves (*V. indicus, V. salvator* and *V. semiremex*). When in freshwater, these species have unlimited quantities of drinking water and therefore have no problems with maintaining water balance. Another species, *V. mitchelli*, also uses river pools, but is primarily arboreal, and frequents trees on the edges of small creeks in northern Australia, occasionally taking to the water to escape capture.

Wild animals will drink water if it is available. This is especially so for herbivores living in hot or arid environments. However, carnivorous animals obtain a large amount of water from their food, as about 70 to 85 per cent of the body of prey animals is available as water. Because of this, carnivores need less drinking water than most herbivores. Goannas have been seen to drink in the wild, although those that live in deserts rarely have the chance to do so. Desert species are frequently subjected to shortages of free water. How do terrestrial and desert varanids cope with these water shortages and with the high levels of salts that they may ingest?

WATER LOSSES

Terrestrial animals lose water to their surroundings in a number of ways. These include evaporation from the lungs during breathing, evaporation from the skin, and loss of water in the urine and faeces. Reptiles, particularly those living in arid environments, show a number of adaptations for restricting their water losses.

EVAPORATIVE WATER LOSS

The amount of water that a goanna loses by evaporation depends on a number of factors. The most important ones are air temperature, movement and humidity, body temperature, skin resistance to water loss and the level and type of activity the animal is engaged in. Laboratory studies have measured the evaporative water loss of *V. rosenbergi* in a dry atmosphere. These show that as the temperature rises from 26° to 38°C the loss of water from the skin and lungs gradually increases by about three times (Figure 8.1). This is because the metabolic rate and respiration rate both increase as the body temperature rises. The air also has more 'drying power' (saturation deficit) at higher temperatures.

When their body temperatures rise above approximately 38°C, goannas begin to flutter the gular pouch under their throat. This causes a dramatic increase in the rate of evaporation from the throat and mouth cavity. Initially, gular fluttering is shallow, at a rate of 60 to 90 movements per minute, with the mouth held open only slightly.

Figure 8.1
Effect of gular fluttering on evaporative water loss in *V. rosenbergi*

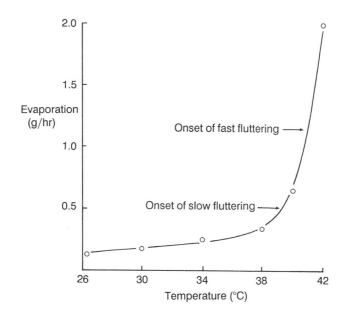

Even so, total evaporation almost doubles. As body temperatures increase, gular fluttering rates increase to more than 120 per minute, and there is a further four-fold increase in total evaporative water loss (Figure 8.1).

Obviously goannas could not sustain such high rates of water loss for long periods of time, as they would rapidly dehydrate. Gular fluttering is a useful short-term method for losing heat from the head and brain, but it is rarely used under natural conditions. In the field, goannas would seek out cooler microclimates, such as burrows or deep shade, to avoid or reduce body heat loads (see Chapter 6, page 59).

The skins of terrestrial reptiles are dry to the touch as they do not contain sweat glands. The skin is thickened with the protein keratin, which, together with lipids, helps to minimise evaporation from the skin (cutaneous water loss). Their rates of water loss through the skin are extremely low compared with those of other terrestrial vertebrates (Table 8.1).

Table 8.1 Rates of water loss from the skin of some lizards and other vertebrates at 30°C

Species	Water loss mg.H_2O/cm²/hr
LIZARDS	
V. rosenbergi	0.12
V. gouldii gouldii	0.11
V. gouldii flavirufus	0.10
V. storri	0.12
V. gilleni	0.06
Sphenomorphus	0.25
Gehyra	0.21
Amphibolurus	0.10
Anolis	0.19
OTHER VERTEBRATES	
Painted quail	0.95
Sheep	1.24 (2.26*)
Man	1.48 (11.6*)
Potoroo	1.7

*While sweating

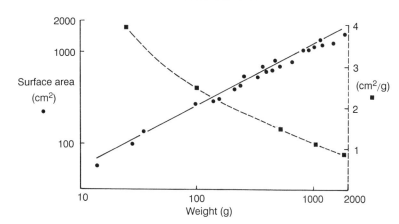

Figure 8.2 Relationship between skin surface area and body weight in *V. rosenbergi*

Evaporation from the skin increases as the temperature rises. *V. rosenbergi* loses 0.12 milligrams of water per square centimetre of its skin at 30°C. The evaporation rate has doubled when its temperature rises to 38°C, due mainly to a higher saturation deficit of the air at higher temperatures (warmer air can carry more water vapour). It is also possible that increased blood circulation to the skin may increase the rate of evaporation.

Large animals have a relatively smaller surface area (square centimetre of skin per gram of body weight) than small animals. The relationship between surface area and body weight in *V. rosenbergi* is shown in Figure 8.2.

The surface area to weight ratio of a hatchling goanna is about five times that of an adult. Because of this, a goanna hatchling would lose far more water across the skin, relative to its body size, than would an adult of the same species. Therefore, juveniles are much more vulnerable to desiccation in hot, dry conditions, and they must behave differently from adults if they are to survive in these circumstances.

About 70 per cent of all evaporative water loss by inactive goannas is through the skin. Another area for evaporative water loss is the surface of the eyes, which are effectively free water surfaces kept wet by secretions of the tear glands (lachrymal glands). Laboratory experiments show that at 30°C in dry air, *V. rosenbergi* loses 67 milligrams of water each hour by evaporation from each square centimetre of eye surface. The eyes of young animals of most species are relatively larger than those of adults, and this is certainly true of goannas. Because of this, relatively more water is lost from the eyes of juveniles than from those of adults. The total eye surface of an adult is only four times that of a juvenile although the animal weighs 100 times more. Thus the water loss per gram of body weight from the eyes of a juvenile is 25 times that of an adult.

THE BIOLOGY OF VARANID LIZARDS

The relative importance of the different avenues of evaporative water loss in goannas of different sizes is shown in Table 8.2. There is comparatively little water loss from the eyes in adults. However, it is interesting to note that they keep their eyes closed while basking, presumably to reduce water losses from this source. In some reptiles, such as snakes and some geckoes, the eyes are covered by transparent scales known as spectacles. These reduce evaporation from the eyes to levels similar to that from the skin. Few desert-adapted reptiles have these 'spectacles', which suggests that 'wet' eye surfaces provide better vision than 'dry' spectacles. The water loss associated with moist eyes can be regarded as an acceptable cost to be paid for having improved vision.

The major sources of water loss in resting goannas are the skin and, in juveniles, the eyes. The respiratory water losses of inactive goannas are amongst the lowest recorded for reptiles. However, once a goanna becomes active within its natural environment, water loss from the lungs increases significantly and becomes the major source of water loss, especially when gular fluttering is used for thermoregulation.

Table 8.2. Effect of size on evaporative water losses of *V. rosenbergi* at 30°C

Weight (g)	Surface area		Evaporation (mg/hr)			Total evaporation (mg. H_2O/g/h)
	Skin (cm^2)	Eyes (mm^2)	Skin+lungs	Eyes	Total (mg. H_2O/g/h)	
15	74	22	8	15	23	1.5
150	341	29	50	20	70	0.5
1500	1611	44	300	29	329	0.2

URINARY EXCRETION

Reptiles produce a number of different nitrogenous waste products that result from the breakdown of proteins. These include ammonia and ammonium salts, urea, uric acid and urate salts. Aquatic species, such as turtles and crocodiles, usually have unlimited access to water to replace that lost during excretion. In these species, the soluble waste products, such as ammonia and urea, predominate. In contrast, terrestrial species generally need to conserve water and cannot afford to excrete waste products that need to be voided with large amounts of water. Apart from one exception, terrestrial reptiles cannot produce a urine more concentrated than the blood. Because of this, the major excretory products in these animals are uric acid and urate salts. These waste products are highly insoluble and require very little water for their excretion.

Birds and mammals are able to produce urine that is much more concentrated than the blood. This is possible because the kidney tubules are folded back on each other (Henle's loops) and there is a concentration gradient across the kidney so that the inner regions have a higher osmotic pressure than the outer regions (Figure 8.3).

The kidneys of reptiles are simple structures compared with those of birds and mammals (Figure 8.3). There are no Loops of Henle and no osmotic gradient across the kidneys of reptiles. No reptilian kidneys are able to produce urine that is more concentrated than blood. One species of agamid lizard (*Amphibolurus maculosus*), living on some salt lakes in South Australia, does produce a urine that is more concentrated than blood. However, the kidneys are not responsible for this. Instead, the urine is concentrated by special modifications to the cloaca and hindgut.

In reptiles the renal artery enters the kidney and subdivides into arterioles that enter the glomerular capsules (or glomeruli). Here, blood pressure forces water, salts and small organic molecules from the bloodstream across to the other side of the capsule membrane. This process is known as glomerular filtration, and the liquid that passes into the kidney tubules is called glomerular filtrate.

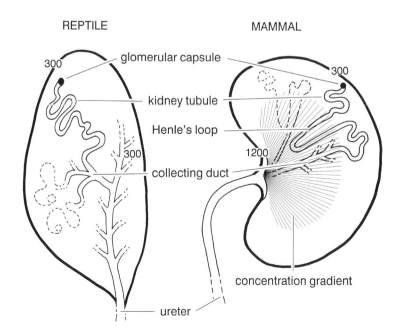

Figure 8.3 Comparative morphology of a reptilian and mammalian kidney (numbers indicate approximate osmotic concentrations in milliosmols)

The filtrate passes down the kidney tubules and, as it does so, water and sodium are resorbed across the walls to re-enter the bloodstream. At the same time, nitrogenous wastes in the form of soluble urates and some uric acid are secreted into the tubules, along with some potassium. By the time the filtrate reaches the ureters it is regarded as urine. Normally, the kidneys of *V. rosenbergi* produce about 7 millilitres of filtrate and 3 millilitres of urine each hour when the goanna's body temperature is 30°C. However, the amount of filtrate and urine produced can vary, depending on the animal's state of water balance and the amount of salt in the body.

Goannas that are dehydrated or burdened with high salt loads reduce the number of kidney tubules in use, reducing the amount of filtrate produced by the kidneys. The animals also resorb more filtrate in the tubules and produce smaller amounts of urine (Figure 8.4). However, if a goanna takes in excessive water, and is water-loaded, the number of working tubules and the volume of filtrate produced are increased. Also, less filtrate is resorbed across the tubule walls. This means that larger quantities of urine are produced and the excess water is excreted. Thus the kidneys are vitally important in maintaining the water balance of the goanna.

In reptiles, changes in the number of working kidney tubules, and the extent to which the tubules resorb filtrate, are regulated by an antidiuretic hormone, arginine vasotocin (AVT). In laboratory experiments, the levels of AVT in the bloodstreams of water-loaded *V. gouldii* are low. Wild *V. gouldii* in summer have more than double the AVT concentration of water-loaded animals, and levels in salt-loaded animals are double that again. There is a clear correlation between circulating levels of antidiuretic hormone and the state of hydration of the animal. When the level of AVT is low, it is likely that all of the glomeruli are working, with only a small amount of filtrate being absorbed from the tubules. This increases urine production (diuresis). At 'normal' AVT levels, the number of functioning glomeruli in the kidney is reduced, and less filtrate and urine are produced. If AVT concentrations are high, as they are in a dehydrated or salt-loaded animal, there is an increase in tubular resorption, resulting in the production of smaller volumes of more concentrated urine.

If high levels of AVT are injected into goannas, even those which are water-loaded, the glomeruli shut down completely, and no urine is produced for some time. When the kidneys commence functioning again, both filtration rate and urine flow are greatly reduced (Figure 8.4). Thus, by means of the action of AVT, stimulated by environmental influences, goannas are able to regulate the volume and concentration of the urine they produce.

Figure 8.4
Kidney function in
V. rosenbergi and
V. gouldii

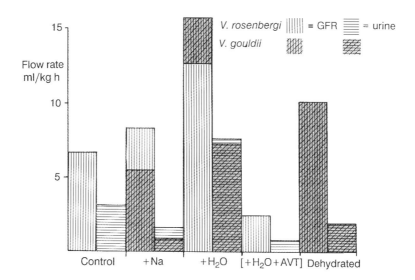

In terrestrial vertebrates the urine leaving the ureters is not immediately expelled from the body. Instead, it is diverted into storage organs such as urinary bladders or cloacas. In reptiles, these storage organs are able to extensively modify the composition of the urine before it is finally excreted.

CLOACAL FUNCTION

The cloaca plays an important role in resorbing water from urine. In most lizards, the cloaca is divided into three main regions. These are the coprodaeum, which is immediately posterior to the hindgut, the urodaeum and the proctodaeum, which opens to the outside (Figure 8.5). The paired ureters open into the cloaca via papillae positioned on the dorsal wall of the urodaeum. Urine emerges from the papillae, and passes forward into the coprodaeum, which plays the major role in water resorption.

Figure 8.5
Cloacal structure
of *V. rosenbergi*
(longitudinal/
vertical section)

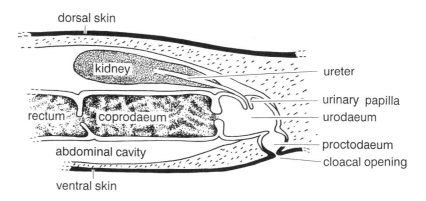

The coprodaeal chamber is the largest of the cloacal regions. Its lining is highly folded, substantially increasing the surface area of the chamber (Figure 8.6). Under a microscope, the surface area is seen to be further increased by small projections, known as villi, which are similar to those found lining the absorptive surfaces of the intestines. At even higher magnification, even smaller projections, called microvilli, are found on some of the surface (epithelial) cells. These give a 'hairy' appearance to the surface of these cells.

The larger the surface area, the greater the amount of resorption that can take place. The folds, villi and microvilli greatly increase the surface area available for the resorption of fluids from the urine. The cloacal wall also has a rich blood supply which ensures the rapid transport of these fluids out of the cloaca.

Cells in the wall of the cloaca acidify the urine when it enters the coprodaeum. This causes soluble urate salts to become insoluble and precipitate, releasing more water to be resorbed.

The cloaca of *V. rosenbergi* resorbs about 8 millilitres of water every hour when the animal is normally hydrated. When a goanna is injected with AVT, it is able to resorb about 21 millilitres of water per hour. Urine in the ureters of goannas is more dilute than the plasma of the blood, even in salt-loaded or dehydrated animals. Because of this, water in the urine can be resorbed across the cloacal wall into the bloodstream by osmosis. This process continues until the urine reaches the same osmotic concentration as the blood. The cloaca can also actively transport sodium out of the urine. This releases even more water for resorption. Eventually, as a result of the precipitation of urate salts and resorbtive processes, the solid wastes form a large pellet of uric acid and insoluble urates within the coprodaeal chamber.

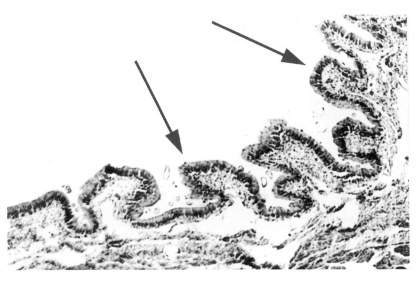

Figure 8.6
Fine structure of coprodaeum (arrows indicate villi)

By the time the urinary pellet is voided from the cloaca, the water content of the pellet is only 48 per cent in normally hydrated goannas. This represents a water loss of only 2.7 millilitres for every gram of nitrogen excreted. This is a much smaller excretory water loss than that achieved by mammals. The most concentrated urine produced by any mammal is that of a desert rodent, *Notomys alexis*, which has an excretory loss of 7 millilitres of water for every gram of waste nitrogen. It is likely that dehydrated goannas produce even drier pellets, further reducing their excretory water losses.

The faeces are formed in the posterior region of the hindgut, immediately anterior to the coprodaeum. This region of the gut has a similar wall structure to that of the cloaca, and resorbs water from the faecal matter. The water content of voided faeces from normally hydrated goannas is 77 per cent. This is much higher than that of the urinary pellets. However, drier faeces are produced when animals become dehydrated.

Like other terrestrial reptiles, goannas have an efficient, flexible excretory system that is well adapted to water conservation. The kidneys produce a dilute urine that results from the tubular resorption of sodium and the limited resorption of water. The kidney tubules secrete soluble urates, which prevent the tubules from becoming blocked by insoluble urate deposits. Once the urine has reached the cloaca, the urates are precipitated to form a pellet, allowing the resorption of further sodium and water. In the unlikely event of a goanna becoming water-loaded, the circulating levels of anti-diuretic hormone (AVT) decline and greater numbers of kidney tubules increase the rate of urine production. AVT levels increase as the goannas become dehydrated, resulting in increased rates of water movement across the kidney tubule and cloacal walls, and increased sodium transport in the cloaca.

Since the resorption of urinary water in both the kidney and the cloaca is dependent upon the active transport of sodium out of the urine, how do goannas, especially those species that enter brackish or even sea water, prevent the build-up of sodium within their bodies?

SALT-SECRETING GLANDS

Sodium build-up is prevented mainly by salt-secreting glands, which are found in many fish, birds and reptiles, especially in marine species. They are also present in many terrestrial lizards, including members of the Varanidae such as *V. rosenbergi*, *V. semiremex*, *V. flavescens*, *V. griseus*, *V. tristis* and *V. salvator*. The salt glands of lizards are in the nasal capsules, those of turtles are lachrymal (behind the eyes), while sea-snakes have sub-lingual and pre-maxillary salt glands in the mouth. It has recently been discovered that some crocodiles have salt glands in the tongue.

Despite the different locations of the salt glands in reptiles, they are all very similar in structure. Each gland consists of densely packed branching secretory tubules radiating from a central duct (Figure 8.7). The central duct connects to the anterior portion of the olfactory chamber and drains to the outside via the nares, eyes or mouth, depending on the location of the gland.

The cells lining the secretory tubules are mainly columnar and have large numbers of mitochondria and lateral cell walls that are arranged into numerous interlocking folds. These structural features are associated with high metabolic activity and the maintenance of an osmotic gradient. These are essential for the production of secretions that are more concentrated than the blood (hyperosmotic).

Herbivorous lizards generally use their salt glands to excrete potassium, which is abundant in their food. On the other hand, marine and some terrestrial carnivorous species, including varanids, secrete mainly sodium, which is more abundant than potassium in their food and environment. At least some species are able to vary their nasal secretions according to whether they ingest high levels of sodium or potassium.

Thus goannas are able to regulate the amount of water and salts in the body by the interplay of the kidneys, cloaca and nasal glands. Urinary pellets are produced that contain little water but large amounts of electrolytes in the form of insoluble urates. Any excess sodium that remains from sodium-linked water resorption in the kidney–cloaca system can be eliminated as concentrated nasal secretions, which again involve the loss of only small amounts of water.

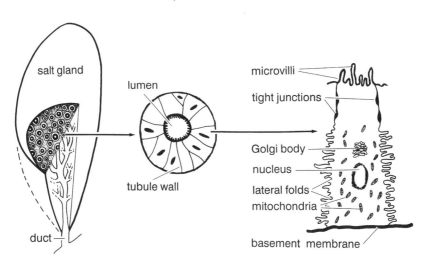

Figure 8.7 Structure of nasal salt gland

WATER INTAKE

Animals gain water from their environments by drinking and by taking in water present in their food. In addition, water is formed when they metabolise food. They also exchange water vapour between the atmosphere and the lung and skin surfaces. The amounts of water that enter (influx) and leave (efflux) an animal's body can be measured by means of heavy (isotopic) water. This is water in which the hydrogen atoms are heavier than normal because of one (in deuterium) or two (in tritium) additional neutrons in the nucleus. While the deuterium atom is stable, tritium is unstable and breaks down in radioactive decay. In the laboratory, it is possible to tell the difference between these different isotopes thus enabling researchers to use them as markers to follow changes in the quantities and the movements of water in the bodies of animals.

If isotopic water is injected into an animal it circulates and mixes with the water in the body tissues and fluids. After a few hours, the isotopic water becomes completely mixed and evenly spread throughout the body. If the precise volume of isotopic water injected is known, then its degree of dilution after mixing gives an accurate measure of the total volume of water in the animal's body. Any water that the injected animal subsequently takes in (by feeding, drinking or vapour exchange) is normal 'unlabelled' water which dilutes the 'labelled' body water. Any water that is lost by evaporation, excretion or vapour exchange is a mixture of both labelled and unlabelled water. Consequently the concentration of labelled water in the animal's body is constantly being diluted (Figure 8.8). Therefore, by measuring the initial dilution of the labelled water and its concentration after a known period of feeding and drinking, the amount of water that entered and left the body during that period (water turnover) can be calculated.

When an animal is in water balance (when water loss and intake are equal), the amount of body water remains constant and a simple equation accurately describes the changes in the water intake of the animal. However, more complex equations are necessary to calculate water exchange if the body water pool changes, as it does with animals that are not in water balance, or are growing.

A number of field studies have been made of the water turnover of varanids living under natural conditions. There are marked seasonal variations in water turnover in two temperate species, *V. rosenbergi* and *V. varius* (Figure 8.9). The lowest daily rates of water flux are found during winter (5 millilitres of water per kilogram), and are about one quarter of the summer values in both species. In winter, the goannas do not feed very often, if at all. The animals are in negative water balance and lose more water than they can obtain. About 60 per

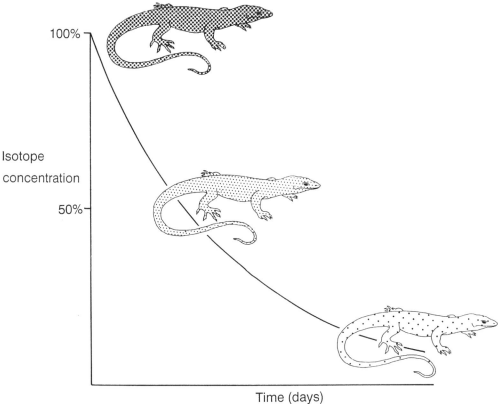

Figure 8.8 Measurement of water turnover by isotope dilution

cent of the water intake is derived from food and the metabolic breakdown of fat reserves and body tissues. The remaining water influx is from the exchange of water vapour between the animal and the atmosphere, both within and outside the burrow, and perhaps by some drinking.

The daily water turnover rates of both species are highest in summer, at about 22 millilitres of water per kilogram. This reflects the greater level of activity of goannas at this time, and their high rates of evaporative and excretory water losses. It also reflects their correspondingly high water intake from increased food consumption, as vapour exchange and drinking are negligible at this time.

The water turnover rates at other times of the year are intermediate between those of summer and winter, those in spring being slightly higher than those in autumn in *V. rosenbergi* (Figure 8.9). *V. rosenbergi* are able to maintain water balance in all seasons but winter.

The seasonal variations in water turnover rates reflect changes in climate, general activity and probably prey availability. The areas over which the animals range also correlate with their water turnover rates.

Goannas forage over much greater areas during summer, when they use most water, than at other times.

Other studies of field water requirements of varanids are not as extensive as those on *V. rosenbergi* and *V. varius* and are usually confined to summer or 'dry' periods. The results of these studies are shown in Table 8.3.

Body size has an important influence on many aspects of an animal's biology, including the surface area (Figure 8.2, page 69) and metabolic rate. As a consequence, it can be expected to affect water turnover. These body-size (allometric) relationships can be described by a general formula indicating the relationship between the body-size of an animal and its rate of water flux. The relationship shows that in mammals and birds, smaller animals generally have higher rates of water flux for each gram of body weight than larger animals. However, in varanids and carnivorous lizards in general, small animals have rates of water flux per gram of body weight similar to large ones (Tables 8.3 and 8.4). The ecological significance of this observation is discussed later in Chapter 9, page 87.

The water turnover rates of arid and temperate region varanids appear to be quite uniform, and are low compared to those of most reptiles. However, in the wet season tropical species appear to have higher rates of water turnover than reptiles from other habitats (Table 8.4). To some extent, this is probably due to these animals drinking, as there is an abundance of free water available in their environments. However, the higher rates of water turnover are also influenced by the higher metabolic rates of these lizards, and the much more humid atmospheres in tropical climes. Both of these factors would increase the exchange of water across the skin and lungs. More research is needed to explain the differences in the water requirements of goannas from different environments.

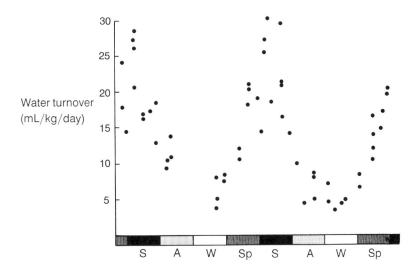

Figure 8.9
Seasonal rates of water turnover in *V. rosenbergi*:
summer (S)
autumn (A)
winter (W)
spring (Sp)

Table 8.3 Water turnover rates of free-living varanid lizards in spring (sp) and summer (su) in arid and semi-arid habitats

Species	Season	Weight (g)	Body water (%)	Water turnover (mL/kg/day)
V. caudolineatus	su	10	79.9	31.6
V. acanthurus	sp	60	70.3	15.9
V. griseus	su	350	77.9	28.3
V. gouldii	su	493	77.0	23.5
V. rosenbergi	su	1,086	76.7	22.0
	sp	1,089	76.5	15.8
V. varius	su	4,300	72.7	24.6
	sp	—	—	15.0
V. giganteus	su	7,700	68.4	22.3
V. komodoensis	su	45,000	74.9	25.5

Table 8.4 Water turnover rates of free-living tropical varanids in the wet (W) and dry (D) seasons

Species	Weight (g)	Body water (%)	Water turnover mL/kg/day	
			W	D
V. scalaris	70	71.8	60.4	16.6
V. mertensi	1,310	70.3	63.2	66.6
V. gouldii	1,090	—	54.7	15.8
V. panoptes	2,400	—	41.8	16.7
V. bengalensis	2,560	76.9	60.5	—
V. salvator	7,600	74.9	54.4	—

CHAPTER 9

ENERGY AND FOOD

THE COST OF LIVING

We have seen that goannas are almost exclusively carnivorous (including insectivory) and prey on a variety of animals to satisfy their energy needs. Animals require energy for all of the physiological processes associated with the normal functioning of the body. These include muscular activity, nervous transmission, growth and reproduction. This energy comes from the process of respiration, that is, the chemical breakdown (oxidation) of glucose, which produces energy, carbon dioxide and water. Glucose is derived from the basic foodstuffs, which are carbohydrates, proteins and fats.

These foods provide different amounts of energy when they are oxidised. Carbohydrates release 17 kilojoules of energy for each gram oxidised, fats yield 39 kilojoules per gram and proteins 24 kilojoules per gram. However, not all of the energy from the breakdown of proteins is available to the animal, as some is lost as nitrogen in the urine. If urea is the main waste product, only 20 kilojoules of energy can be gained by an animal from the metabolism of each gram of protein, whereas if urates or uric acid are excreted, only 18 kilojoules can be gained from each gram of protein metabolised.

Terrestrial animals all contain relatively high levels of water and protein, and reasonable amounts of fat. Thus a carnivore can obtain large amounts of water and energy from its food. In contrast, plants vary extensively in their content of water, protein, carbohydrates and energy (Table 9.1). Grasses, in particular, can be extremely dry and fibrous during drought and yet they contain high levels of water after good rains. Because of this variation in plant water content, carnivores are much less affected by changes in environmental and climatic conditions than are herbivores. Carnivores are what they eat.

Not all of the prey material of a carnivore can be digested. Fur, feathers, some bone and invertebrate exoskeletons pass through the gut and are eliminated in the faeces. Generally between 80 and 90 per cent of the total energy present in prey can be digested and metabolised by carnivores such as goannas. Herbivores feeding on fibrous diets cannot obtain as much energy from their food as carnivores.

When food materials are metabolised, they require different amounts of oxygen, and produce different amounts of carbon dioxide. Animals that are metabolising only carbohydrate produce as much carbon dioxide as the amount of oxygen used. However, if an animal is metabolising only fats, as happens during long-term fasting, carbon dioxide production is only 70 per cent of oxygen consumption. For protein-based metabolism, carbon dioxide production is about 80 per cent of oxygen consumption. Carnivore diets provide very little carbohydrate, as the main food components are protein and some fat. This type of diet would produce carbon dioxide volumes of about 75 per cent of the oxygen used.

It is quite easy to measure the amount of carbon dioxide that an animal produces in the laboratory, or the amount of energy it uses (its metabolic rate) when it is kept under artificial conditions. However, only in recent years has a method been developed to measure accurately the metabolic rate (carbon dioxide production) of free-living animals. This method is commonly called the doubly-labelled water turnover technique and is similar to the isotopic hydrogen water turnover technique described in Chapter 8, page 77. The technique uses water labelled with an isotope of oxygen (^{18}O) as well as water labelled with tritium (^{3}H) or deuterium (^{2}H). ^{18}O has a mass of 18 instead of the usual mass of 16 for oxygen. When ^{18}O is injected into an animal in labelled water, it is lost from the body not only in the form of water, but also as respiratory carbon dioxide produced during metabolism. By injecting an animal with both hydrogen-labelled and oxygen-labelled water it is possible to measure the rate of water loss, the amount of ^{18}O isotope lost as water and the total amount of ^{18}O

lost. The difference between the total ^{18}O loss and that lost in water is the ^{18}O lost as carbon dioxide. The relative loss of isotopes is shown diagrammatically in Figure 9.1.

The doubly-labelled water method has been used to study the field metabolic rates and energy use of a wide range of animals, including a number of varanids.

The most detailed study of energy acquisition and use by a goanna has been made on *V. rosenbergi*. The seasonal pattern of carbon dioxide production is similar to that of water intake. The highest metabolic rates are in summer, the lowest in winter, and they are intermediate in spring. Metabolic rates were not measured in autumn. The amount of food a goanna requires to balance its metabolic energy expenditure can be calculated if the energy content of

Table 9.1 Composition of some dietary items

Species	Water (% of weight)	Energy (kJ/g fresh weight)
VERTEBRATES		
Leilopisma sp. (skink)	73.6	5.16
Gehyra variegate (gecko)	74.0	5.31
Mus domesticus (house mouse)	65.9	7.57
Macropus eugenii (tammar wallaby)	69.6	6.46
INVERTEBRATES		
Insect larvae	79.1	5.02
Isopods	64.9	5.05
Beetles	65.4	8.61
Earthworms	75.1	3.53
Centipedes	71.9	6.11
Scorpions	68.9	7.54
Cockroaches	70.4	5.67
PLANT MATERIAL		
Dry grass	14.0	16.28
Lush grass	84.0	3.03
Gum leaves	57.0	9.46
Saltbush	82.0	3.75

Note: Generally 80 to 90 per cent of the energy content of animals can be digested and assimilated by predators, however herbivores can usually obtain only approximately 50 per cent of the energy contained in grass and leaves.

the diet is known. In making such calculations, it can be assumed that 26 kilojoules of energy are expended for each litre of carbon dioxide produced, and that 85 per cent of the energy content of prey is digested and metabolised.

These calculations show that in summer an average *V. rosenbergi* weighing 1 kilogram uses 112 kilojoules of energy each day and would need to consume about 22 grams of food to balance its energy expenses. In winter the same goanna would expend only 22 kilojoules each day and need only 4.5 grams of food to achieve energy balance. Even so, a goanna does not feed regularly in winter and uses more energy than it gains for much of this period. The extra energy needed in winter comes from fat stored in the body during summer. Since the water influx rates in spring and autumn are similar in *V. rosenbergi*, it can be assumed that the metabolic rates in spring and autumn are also similar. We can therefore assume that a 1 kilogram goanna will use 62 kilojoules per day, and consume about 12 grams of prey each day during these seasons.

A total annual energy budget for *V. rosenbergi* is shown in Table 9.2. For these calculations, it was assumed that summer and winter each last 120 days, while spring and autumn each last 60 days. Over a whole year, a 1 kilogram goanna will use 23 600 kilojoules of energy and will need to eat about 4700 grams of food to balance its energy expenditures.

This annual energy budget assumes that the goanna neither gains nor loses weight. If an animal is still growing, laying down fat reserves or producing eggs, it will have a higher energy expenditure and food

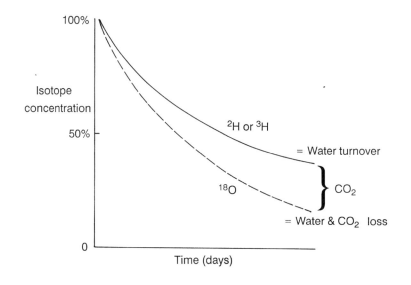

Figure 9.1 Measurement of carbon dioxide production using doubly-labelled water

Table 9.2 Annual energy budget of a 1kg V. rosenbergi

Season	Days	Energy spent (kJ/day)	Seasonal total Energy (kJ)	Food (g)
Summer	120	111.7	3,400	2,680
Autumn	61	62.4	3,800	760
Winter	123	22.5	2,800	550
Spring	61	62.4	3,800	760
Annual total			23,800	4,750

requirement. We can calculate the total amount of energy stored in an average clutch of eggs as follows. The average clutch size of *V. rosenbergi* is 14 and each egg weighs an average of 25.7 grams. The average energy content of an egg is 8.1 kilojoules per gram, so the total energy content of a clutch is 2900 kilojoules. This is about 12 per cent of the total energy used in a year by a non-breeding goanna. However, a lean female entering the breeding season in early spring has only four months in which to accumulate the energy needed to produce a clutch of eggs. Therefore, it is more realistic to compare the energy stored in the clutch with the daily energy expenditure during spring and early summer. When this is done it is found that the energy commitment in a clutch represents an increase of 32 per cent over the average daily energy requirement during the breeding season. If the biochemical energy cost involved in egg production in goannas is similar to that of other lizards, then egg production represents a 40 per cent increase over non-breeding energy expenditure.

To acquire the extra energy needed to produce a clutch of eggs, a female would need to capture an additional 668 grams of prey during spring and early summer. If prey or carrion are not relatively abundant during this time, it may be difficult for her to accumulate adequate energy reserves to enable her to breed successfully every year. However, if a female does not accumulate enough energy to breed in one year, she may enter winter with a greater than normal energy reserve, and this may increase her chances of breeding successfully the following year. Interestingly, some captive varanids are able to produce two clutches of eggs in a year, probably because of their abundant and regular food supply.

Goannas that acquire more energy than they need lay down

reserves of body fat in a pair of fat bodies. These are located in the posterior region of the abdominal cavity (Figure 3.6, page 24). In lean animals, these fat bodies are small, but in some animals they can reach a substantial size. They can weigh at least 13.7 per cent of the total body weight in *V. bengalensis*, 10 per cent in *V. giganteus* and 7.6 per cent in *V. rosenbergi*. Some species, including *V. rosenbergi* and *V. gouldii*, also store some fat in the tail.

The field metabolic rates (FMRs) of hatchling *V. rosenbergi* are about five times higher than those of adults in spring. This is due to a number of factors: the standard resting metabolism of hatchlings is higher, the young are actively growing, and they maintain a high overnight body temperature due to the occupation of termite mounds. Experiments have shown that this latter influence is the most important. Hatchlings that are maintained away from termite mounds have field metabolic rates that are similar to adults.

The FMRs of some varanids in different seasons are shown in Table 9.3. The pattern of energy use parallels that of water flux; rates are highest and lowest for tropical species in the wet and dry seasons respectively, with FMRs of arid/semi-arid species in summer being intermediate.

Table 9.3 Field metabolic rates of arid/semi-arid and tropical varanids in spring (Sp), summer (S), wet (W) and dry (Dry) seasons

Species	Season	Mass (g)	CO_2 produced (mL g^{-1} h^{-1})	Energy used (kJ kg^{-1} d^{-1})
V. caudolineatus	S	10	0.46	—
V. acanthurus	Sp	60	0.10	63
V. rosenbergi	Sp	1089	0.10	59
	S	1086	0.18	114
V. giganteus	S	7700	0.17	101
V. scalaris	W	66	0.21	128
	D	77	0.11	68
V. mertensi	W	1208	0.20	121
	D	1130	0.13	81
V. gouldii	W	1150	0.31	195
	D	1055	0.11	69
V. panoptes	W	2300	0.22	136
	D	1550	0.09	54
V. bengalensis	W	2560	0.25	156
V. salvator	W	7600	0.20	125

ENERGY AND FOOD

A comparison of the annual energy requirements of some free-living vertebrates similar in size to a Rosenberg's goanna (about 1 kilogram body weight), and estimates of their food and water intakes, is shown in Table 9.2.

It is clear that the goanna is far more conservative in its use of energy and water than are 'warm-blooded' (homeothermic) mammals and birds. A goanna eats about five times its own body weight of prey each year. By comparison, a marsupial carnivore, such as the eastern quoll (*Dasyurus viverrinus*) eats about 65 times its own weight, a marsupial herbivore, such as the brush-tailed bettong (*Bettongia penicillata*) about 70 times and a bird like the little penguin (*Eudyptula minor*)

Table 9.2 Comparative use of water and energy in similar sized (1kg) reptile, mammals and a bird

	Energy (MJ)	Food (kg)	Water (Litres)
Rosenberg's goanna (*Varanus rosenbergi*)	24	4.7	5
Eastern quoll (*Dasyurus viverrinus*)	327	65.4	56
Brush-tailed bettong (*Bettongia penicillata*)	212	68.9	42
Little penguin (*Eudyptula minor*)	446	115	108

about 115 times its own weight annually. The goanna makes substantial savings in water and energy expenditure by allowing its body temperature to drop during the night, and by remaining relatively inactive and at a low body temperature during winter. The homeotherms pay a heavy price in energy use in order to stay 'warm-blooded'.

Most reptiles show characteristically low field metabolic rates and water turnover rates. This makes them ideally adapted to exploiting habitats lacking in energy and water, such as deserts. Ectotherms, such as varanids, are very successful in such harsh environments. Their ability to allow major variations in body temperature to occur, and their ability to remain inactive for long periods, should not be regarded as 'primitive' or 'inferior'. Instead, such strategies should be looked upon as effective alternatives to those used by homeotherms in the use of water and energy.

The majority of lizards are smaller than most mammals and birds. Eighty per cent of species weigh less than 20 grams and 65 per cent of species less than 10 grams. It seems that body size does not influence the water, energy and food requirements of free-living varanids to the same extent that it does in birds and mammals. Because of this, lizards do not incur the penalties that small size imposes on homeotherms such as shrews or hummingbirds, which need great amounts of energy to remain active. Thus lizards can be very small and are able to exploit ecological niches that birds and mammals cannot use because of their high energy and food requirements.

PARASITES

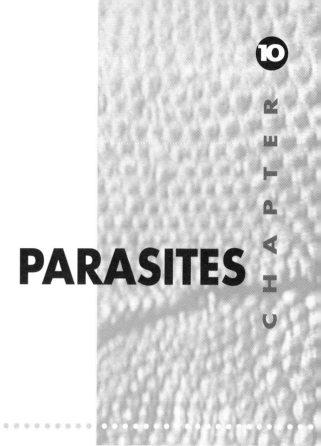

Varanids, like most other vertebrates, have a number of other animals living on them (ectoparasites) or in them (endoparasites). These are mainly ticks or gastric nematodes.

ECTOPARASITES

The ticks which occur on varanids in Australia, southeast Asia and Africa are hard ticks, mainly in two genera (*Amblyomma* and *Aponomma*) which are in the family Ixodidae. These genera occur on a wide range of vertebrates, including other reptiles, particularly snakes and turtles, and birds, monotremes, marsupials and occasionally placental mammals, including humans.

There are three major species of each genus which occur on varanids in Australia: *Amblyomma limbatum*, *Am. albolimbatum* and *Am. moreliae* and *Aponomma fimbriatum*, *Ap. undatum* and *Ap. hydrosauri*. The major species on varanids in Asia are *Amblyomma helvolum* and *Aponomma trimaculatum* and in Africa they are mainly *Aponomma flavomaculatum*. However, there are also at least 20 species of ticks in these two genera which are known to infest varanids (Table 10.1). Some of these species, such as *Amblyomma glauerti*, *Am. perenticola*, *Am. calabyi* and *Am. robinsoni* and

Aponomma glebopalma, Ap. soambawensis, Ap. latum, Ap. varanensis and *Ap. komodoense*, infest only one or two species of varanids and thus are very host-specific while others (*Amblyomma limbatum* and *Am. helvolum* and *Aponomma undatum* and *Ap. trimaculatum*) are known to infest at least six varanid species and are thus generalists (Table 10.1).

A number of different species of ticks may be found on some species of *Varanus*: four species (*Am limbatum, Ap fimbriatum, Ap. undatum* and *Ap. hydrosauri*) have been recorded on *V. gouldii*, and three (*Am. moreliae, Ap fimbriatum* and *Ap. undatum*) on *V. varius*. Some species of varanids can have different species of ticks in different parts of the range of the host species. The highly host-specific species generally have very restricted distributions.

Not all species of varanids in an area are infested by all the species of reptile ticks in the area. While *Aponomma gervaisi* is frequently found on *V. bengalensis* in India and Pakistan (87 per cent infested), in areas where *V. flavescens* and *V. griseus* also occur, it did not occur on any specimens in the large samples of these other two species which were examined.

Both *Amblyomma* and *Aponomma* are three-stage parasites and develop from eggs, which are deposited in the soil, and pass through larval, nymph and adult stages, each requiring attachment to a separate host whose body fluids they suck while they develop. These larval and nymphal stages have highly variable periods during which they complete their development, which can take up to three years. Most species of ticks can only locate their hosts over a very short distance, and some species position themselves on vegetation and wait for potential hosts to pass by and brush against them. They may also locate hosts by encountering them in refuge sites, but most spend long periods off their host species. Moving about in search of hosts would require the use of large quantities of water and energy. Waiting for potential hosts in areas which they use frequently is thus far more economical.

There is one report of a tick found beneath the skin (sub-dermal) of a varanid and anecdotal accounts of ticks attached to the walls of the cloaca. All three on-host stages of ticks (larvae, nymphs and adults) can occur on a host at the same time. The greatest number of ticks are generally found on large individuals of the larger species of varanids (Table 10.1).

Different species, sexes and stages of ticks attach at different sites but these are generally in areas of soft tissue, such as in the nares, axils of the legs, between the toes or scales, or around the cloaca. The attachment sites selected by adult female ticks in the species

Table 10.1 Tick species infesting varanid lizards

Host species	Host number	No. with ticks	% infested	Host SVL* (mm)	Tick species
Australian species					
V. caudolineatus	156	0	0	137	
V. gilleni	21	2	10	159	Amblyomma sp.
V. eremius	65	2	3	165	Amblyomma sp.
V. pilbarensis	10	1	10	169	Amblyomma sp.
V. glauerti	42	15	36	227	Amblyomma glauerti
					Aponomma glebopalma
V. acanthurus	127	6	5	237	Amblyomma limbatum
V. mitchelli	10	2	20	247	Amblyomma limbatum
V. scalaris	102	6	6	253	Amblyomma limbatum
V. tristis	62	10	16	280	Amblyomma limbatum
V. glebopalma	44	32	73	335	Amblyomma glauerti
					Aponomma glebopalma
V. rosenbergi	46	39	85	395	Aponomma fimbriatum
V. panoptes	23	6	26	500	Amblyomma limbatum
V. giganteus	14	3	21	700	Amblyomma calabyi
					Aponomma sp.
Asian and African species					
V. timorensis	47	17	36	185	Aponomma soambawensis
V. indicus	20	12	60	365	Aponomma trimaculatum
V. griseus	3	2	67	430	Aponomma gervaisi
	40	30	75		Aponomma varanensis
	76	2	2		Aponomma laeve
	51	4	8		Amblyomma helvolum
V. niloticus	112	104	93	575	Aponomma flavomaculatum
V. olivaceus	116	111	96	~600	Amblyomma helvolum
	116	3	2		Aponomma fimbriatum
V. bengalensis	70	63	90	610	Aponomma gervaisi
V. salvator	91	65	72	1040	Amblyomma helvolum
V. komodoensis	?	?	?	1340	Amblyomma helvolum
					Amblyomma robinsoni
					Amblyomma komodoense

*SVL = snout–vent length

Aponomma gervaisi are generally on the anterior portion of the host in protected thin-skinned areas which are heavily vascularised, as their diet consists almost entirely of blood, which is highly nutritious. Attachment sites of males of that species are generally in the posterior areas of the host, near the cloaca or on the tail. They break down cells in these regions and feed on the digested cellular material, fluids and lymphoid cells. This is also the probable reason for clusters of males attaching themselves in the regions of wounds.

At least some, if not all, species of ticks mate on the host species. Mating and subsequent engorgement of female ticks on reptiles only occurs during the warmer months when their hosts are most active. Females can increase in size by up to five-fold following engorgement. They then leave the host and oviposit in the ground. Although the percentage of hosts infested with male or female ticks does not differ, the number of males of most species of ticks found on reptiles are usually far greater than the number of females, as males can remain on a host for several months whereas females remain on the host for much shorter periods. The males are more active in their search for mates than the females, and they probably locate receptive females by means of airborne pheromones which are transported along the host's body as it moves.

There are substantial differences between tick species in many aspects of their life history, such as the time taken to complete different stages of development, the size and number of eggs produced and the ability of eggs and other developmental stages to withstand desiccation and thus the length of time that they can spend off their hosts. These parameters determine which species are found in particular habitats.

It is not known whether ticks have a detrimental effect on varanids (such as by lowering their longevity, growth rate, or influencing their reproductive success), but tick loads appear to have little or no effect on these factors in the skink *Tiliqua rugosa*. One possible negative effect ticks may have on reptiles may be in the transmission of diseases. For instance, some reptiles, including several species of varanids, carry malarial plasmodia which may be transmitted by ticks.

ENDOPARASITES

The most common and abundant endoparasites of varanid lizards are nematodes in the genus *Abbreviata*, which also occur in many species of snakes. These are generally ingested indirectly in prey species which are intermediate hosts (i.e. host capture), ranging from invertebrates to reptiles, birds or mammals. Termites appear to be particularly important as intermediate hosts in arid regions of Australia. Few species of lizards or elapid snakes in Australia are not hosts to adult or larval nematodes. Nematodes of the genus *Tanqua* are also found frequently in several Australian and Asian species of *Varanus*, particularly those associated with aquatic habitats. Other nematodes occasionally found in varanids include *Amplicaecum, Ambiostrongylus, Hastaspiculum, Skrjabinoptera, Maxvachonia, Dioctowittus, Wanaristrongylus, Meteterakis, Physalopteroides* and *Oswaldofilaria*. Some of these genera are not found in Australian varanids.

All these nematodes develop from eggs and pass through four larval stages. Little is known of the intermediate life stages of gastric nematodes in reptiles but many species of invertebrates, lizards and snakes are involved in them. Many small reptiles only contain larval stages of nematodes, often encysted in the walls of their stomachs, which do not mature until they are consumed by large species of reptiles. Some gecko species have a 100 per cent prevalence of nematode larvae and they are usually species which mainly consume termites.

Some species of *Abbreviata*, such as *A. perenticola* or *A. glebopalmae*, have only one or two host species whereas others, such as *Abbreviata hastaspicula, A. confusa* or *A. levicauda*, have several varanid species as hosts. Most individual varanids carry only one species of nematode, but widely distributed species may be parasitised by several different but closely related species of nematodes in different parts of the host's range, and some individuals carry as many as six species. Several species of varanids (*V. indicus, V. timorensis, V. mertensi* and *V. scalaris*) have a low prevalence (number of infected individuals) of adult nematodes, but most species have a prevalence of 46 to 100 per cent. Small species of varanids have a very low prevalence of adult nematodes, and none have been reported to occur in *V. komodoensis*.

Host species may have different densities of nematode species depending on the habitat they utilise. The prevalence and intensity of infestation (number of nematodes in an individual host) differ greatly within and between species, possibly as a result of dietary and habitat differences. In addition, the larger and thus probably older individuals generally have the highest intensity of infestation, most probably because they have had a longer period to acquire parasites and because they eat more food than smaller individuals. There may also be seasonal fluctuations in the intensity of infestation of varanids. The greatest intensity of infestation occurs in large varanids in arid regions of Australia, probably as a result of their prey species feeding predominantly on termites containing intermediate stages of nematodes. They are unusual in this regard as the desert reptile community in Australia generally supports an attenuated fauna of gastric nematodes, with a small number of species and generally low host specificity and low intensity of infestation. The greatest diversity of nematode species found in varanids occurs in tropical areas in northern Australia where there is a high diversity of habitats, although intensity of infestation is generally low in that area. The composition of the nematode community in a host species appears to be a result of diet rather than phylogenetic relationship.

Gastric nematodes feed on the stomach contents of their hosts. When there is no food in the stomach, some species of nematodes attach loosely to the stomach walls. Nothing is known about the effect nematodes have on their hosts, but it seems likely that they utilise energy by consuming their host's food, and thus reduce the energy available to the hosts.

Other types of endoparasites which occur in varanids include cestodes, trematodes, pentostomatids and protozoa. Little is known about cestodes in varanids; there are several species in the genera *Acanthatenia, Dutherisia* and *Scysocephalus* which are generally found in Asian or African species. Two other genera of tapeworms (both Protocephalidea) are known to occasionally infest Australian varanids.

There are two genera of pentostomatids (*Elenia* and *Sambonia*) known to infest varanids in Australia and Africa respectively. *Elenia* have three stages of development in mammals, snakes and varanids, and adults occur in mammals and varanids in Australia. Adults infest the lungs or the mouth and tongue region of reptiles.

Malarial plasmodia (*Eimeria, Entamoeba, Plasmodium* and *Endolimax*) have been detected in several Asian species (*V. prasinus, V. salvator, V. bengalensis* and *V. komodoensis*), but their effect on these host species is unknown. Blood protozoans of the genera *Haemogregarina, Hepatozoon, Sarcocystis* and *Cryptosporidium* have been identified in four species of Australian varanids (*V. giganteus, V. gouldii, V. tristis* and *V. varius*).

While many species of *Varanus* carry large numbers of parasites, they seem to suffer little ill effect from doing so. There is some host specificity in these parasites, but some species are generalists and often the biogeographic distributions of hosts and parasites do not coincide. In many cases the parasites will also utilise species other than varanids as their final host, and several species of varanids will be hosts to different species of parasites in different parts of their distribution.

And just to show how fair it all can be, some of the parasites of varanids can be parasitised in turn by other parasites. A chalcid mite, *Hutterella hooker*, is commonly found parasitising ticks in India, including the varanid tick, *Aponomma gervaisi*. But then, why not?

CONSERVATION AND MANAGEMENT

Many reptile species are hunted and exploited for commercial purposes by humans, and varanids are no exception. Humans are also responsible for many perturbations of natural habitats such as land clearing and pollution, which also exert pressures on populations of varanids. Let us look at the conservation status of varanids on a regional basis.

AUSTRALIA

The conservation status of all Australian varanids is sound. They are protected from exploitation by state, federal and international legislation which is generally well enforced. Commercial use for skins or meat is prohibited and collection or keeping of varanids is permitted only for research purposes or by approved zoos in some states, while others are more liberal in who they allow to keep varanids.

Several species of goannas are used for food by Australian Aborigines, but the extent of their use and the species involved vary from region to region. They comprise only a small part of the diet of residents of coastal regions, but for desert dwellers they may make up a larger proportion of the meat in the diet than large game such as kangaroos and emus. The survival of goannas in any area is probably not threatened by Aboriginal exploitation, as hunting for them is generally done on an opportunistic basis. Goannas are occasionally collected when encountered by the hunters who are searching for large game, but most are collected by women and children who are foraging for plants and small animals.

Goannas feature in many of the dreamtime stories of the Aborigines and are important dreamtime figures believed to occupy some of their sacred sites. They are represented in many of their rock and bark paintings, and feature as tribal and personal totems. They are thus very important animals in many Aboriginal cultures.

Some goannas are killed by vehicle traffic on roads, but roads and heavy traffic are uncommon in most parts of Australia and most goannas do not encounter this problem. It may, however, become more of a problem as traffic increases in national parks and other areas where varanids are abundant.

Habitat alteration does pose a threat to some local populations. Highly developed urban areas obviously provide little or no suitable habitat for goannas. The larger species of goannas in southern Australia are generally still abundant in patches of remnant native vegetation in areas where farming has been carried out for many years and has greatly altered the natural habitat. These species are also often seen in the outer metropolitan areas of even the larger cities.

Most species of Australian goannas have wide distributions so that habitat alterations are unlikely to occur over the entire range of the species. There are some species with restricted distributions, such as *V. kingorum* and *V. pilbarensis*, but very little research has been done on either of these species and the actual distributions may be larger than are currently known. Also, both of these species exist in remote areas where the level of human activity is likely to remain low.

A different kind of indirect human impact that poses a major threat to many species of goannas in northern Australia results from the presence of the cane toad (*Bufo marinus*). It was introduced into Queensland from South America, via Hawaii, in 1935 in a misguided attempt to control insects in sugar-cane fields. It failed to do so, and has now spread over much of wet-tropical northeastern Australia. It is predicted that it may reach Darwin and the vast wetlands of northern Australia by approximately the year 2002. The toads secrete a highly toxic substance from glands located in the skin on the shoulders. This poison is fatal to large varanids and other predators which attempt to eat the toads. The toads themselves are also large animals and voracious feeders, which eat almost any type of prey. They are believed to have already had a substantial impact on the small vertebrate fauna in large areas of Queensland, and may compete with varanids for food.

CONSERVATION AND MANAGEMENT

One of the few commercial uses of goannas by European settlers in Australia was the production of the patent medicine 'Goanna Oil', which supposedly was able to cure a wide range of ailments. These appeared to have included almost any ailment its users wished to treat. It was believed to possess amazing properties, including the oft-repeated claim that the oil was so fine that eventually it would penetrate the glass bottle in which it was kept. In reality, the oil was a mixture of wintergreen, menthol, pine, peppermint and eucalyptus oils. Originally it did contain oil from fat bodies of *V. varius* but for many years has contained no oil from goannas. It is unlikely that enough goannas were ever used in its production to have had any significant effect on their numbers, even within the local area near the factory.

ASIA

The conservation status of several species of monitors in southeast Asia is strikingly different from that of Australian species. In most countries in the region there is a long tradition of exploiting monitors for food, use in traditional medicines and for skins which are used locally in the manufacture of drums and other objects. In India, there are cave paintings which are over 10 000 years old which appear to depict monitors as food. Varanid eggs are also highly regarded as food in many areas, but this exploitation probably seriously affects only monitor populations in heavily settled areas.

Since skins have become commercially valuable, and highly so in the under-developed areas that comprise most of southeast Asia, huge numbers of several species of monitors (*V. salvator, V. bengalensis, V. indicus* and *V. flavescens*) have been killed to provide skins for the fashion industries of wealthy countries. Little or nothing is known of the population sizes or dynamics of any of these species. The maximum sustainable rates of harvest are unknown, so it is not known whether the numbers being harvested are acceptable or not. The accuracy of the information on the actual numbers being killed is also not known. Live animals or their skins may be easily shipped across national borders, so international laws must be enforced to limit this trade.

Available figures show that *V. salvator* is the Asian species of the greatest commercial importance. Over 220 000 skins of this species were exported from ten Asian countries in 1980. This number did not include over 175 000 skins of *V. salvator* which were exported from countries where the species does not even occur. The number of skins exported from the same ten Asian countries during the following five years ranged from 425 000 to 1 300 000, and those from the countries where they do not occur or from unknown sources ranged from 52 000 to 223 000. The main source of skins was Indonesia, which exported almost 1 800 000 during the entire period. Over 750 000 were exported from Singapore, which has a total area of 600 square kilometres. Obviously the vast majority of these skins must have originated in other countries. The main countries which imported these skins were Belgium, France, Italy and Japan. There is often poor agreement between the number of skins supposedly exported and those reported as being imported.

The number of *V. salvator* skins which have been exported from Indonesia varies greatly. There were 83 000 skins exported in 1920, which rose to 650 000 in 1926. Most of these came from Java, which recorded exports of 400 000 in 1926. Skin exports began to decline in the next decade, as shown by those originating in Borneo in 1933 (125 000), 1934 (36 500) and 1935 (35 800). No further data are available until 1980 when the export of 81 000 skins was reported and the numbers have increased steadily since then to 700 000 in 1987. The number of skins harvested appears to have remained at approximately that level until 1993. Most of the animals were caught by semi-professional hunters who are usually farmers or fishermen. The harvest mostly consists of animals in the age range of three to six years, which means they have little opportunity to breed.

Extensive hunting of *V. salvator* still occurs in the Philippines despite the fact that the export of wildlife products is supposedly subject to strict control. In 1985, over 80 000 skins from the island of Mindanao alone were exported to Japan. Many of these were the brightly coloured *V. salvator cumingi* (Plate 13). Populations of this species are decreasing in many parts of the country because of deforestation and clearing of mangrove areas. Hunting for skins, which occurs even in wildlife sanctuaries and other sparsely inhabited regions, exacerbates the situation. *V. salvator* is the species which is most exploited for skins and meat. It does, however, have the largest distribution of any varanid species.

The only other Asian species for which export figures are available is *V. indicus* In a ten-year period, a maximum of 13 000 skins were exported (mainly from Indonesia) in any year. This species is very widely distributed in the south Pacific region, where it was apparently spread deliberately by German, Japanese and American troops in the twentieth century as a potential source of food, and as an intended pest control agent. It has, however, become a pest itself because of its predation on fowl and other beneficial animals, and is often killed for this reason. In some areas it is also used as food.

The skins of other Asian species are used by leather manufacturers but the numbers that are harvested are apparently much lower than those of *V. salvator*, or they are not reported. Populations of

V. bengalensis in Sri Lanka were reported to be heavily reduced in the early 1950s because they were hunted for skins, but this trade no longer exists. In India, monitors (*V. bengalensis*, *V. flavescens* and *V. griseus*) are all legally protected but they are still caught and used for many purposes and their numbers have greatly declined in most areas.

Similar trends have occurred in the harvest rates for skins of other large reptiles (lizards, snakes and crocodiles). However, the largest living lizard, *V. komodoensis*, is not hunted for its skin as the scales contain large bony inclusions (osteoderms) that make the skins unsuitable for manufacture into leather goods. In addition, *V. komodoensis* occurs mainly in a national park, where it is legally protected. However, most of the exploited wildlife species have had legislative controls and restrictions placed on their exploitation though it is often impossible to enforce this legislation. Inadequate numbers of enforcement officers, lack of equipment, official disinterest or corruption, poor education levels of the local people (and often of customs officials) and the high economic value of the skins can all contribute to ineffectual legal enforcement. If, as seems likely, the high number of monitors being harvested is excessive, legal constraints seem unlikely to provide the necessary level of protection. In addition, the level of legislative protection of monitors varies greatly between the countries of southeast Asia.

The most successful means of reducing the number of goannas killed would appear to be reduction of demand, and thus the market value of the skins. This could be achieved by persuading the potential buyers of skin-based goods such as luggage, handbags, shoes, belts and watchbands that such purchases are ethically unacceptable. This method has proved successful in reducing the demands for other products such as furs and ivory, which when harvested in a largely uncontrolled fashion can endanger the survival of particular species. Another useful approach is the development of well-designed farming programs to produce skins and thus minimise the threat to wild varanid populations. Commercial crocodile farms are already operating successfully in a number of countries. Captive monitors have also been bred successfully in recent years at a crocodile farm in India. Both male and female *V. salvator*, which have been held in captivity in Indonesia, reached breeding condition when they were two years old and captive females have produced clutches of 1 to 17 eggs.

Another important factor which could have a major impact on monitors in southeast Asia is the massive habitat destruction occurring over much of the region. Forest-dwelling species are at particular risk from this practice, as huge areas are being cleared by logging, either legally or illegally. Additional clearing is creating more agricultural land, often in unsuitable areas. The large and rapidly increasing human populations of the region are exploiting the natural resources at an increased rate, and there is a continual demand for more land. Undeveloped areas are under increasing threat. The conservation ethic is either a totally alien concept or is of little or no concern to poor peasants when they can see opportunities to obtain more food, money or land.

The effects of pollution by industrial wastes or agricultural chemicals on monitors in this region are unknown. Far fewer restrictions on industrial pollution exist in most parts of southeast Asia than in more affluent regions of the world. Many agricultural pesticides that have been banned from use in Europe or North America are still readily available in most southeast Asian countries. Few data are available on their influence on the health or reproductive performance of monitor populations in those countries, or on other animal or human populations. Because monitors are carnivores and are near the top of the food web, they probably concentrate and accumulate within their tissues many toxins derived from their prey .

AFRICA

The desert monitor *V. griseus* is widespread over most of North Africa, but is at very low densities in many areas due to very harsh desert conditions. Most of the area also has a very low human population, which is largely nomadic. The nomads use many parts of the monitor in their traditional medicines. The Nile monitor (*V. niloticus*), the white-throated monitor (*V. albigularis*) and the savannah monitor (*V. exanthematicus*) have long been harvested for their skins, for food and for use in traditional medicines.

The high commercial value of monitor skins has increased the incentive to hunt them. The total number of *V. niloticus* skins exported from 11 countries between 1980 and 1985 was 2 773 000, and varied from 263 000 skins in 1983 to 566 000 in 1982. Between 1985 and 1990 an average of 465 000 (and in one year a maximum of 700 000) skins of this species were exported. Up to 210 000 skins of this species were exported in one year from unknown sources and from countries in which *V. niloticus* does not occur. Most skins were exported from Sudan, Cameroon, Chad, Nigeria and Mali. The main countries that imported them were Italy, France, Switzerland and the United States. The majority of these animals were caught on baited hooks.

Fewer *V. albigularis* and *V. exanthematicus* skins were exported than *V. niloticus* skins. The maximum number exported during any year between 1980 and 1985 was 240 000 in 1981, and the minimum was 14 000 in 1984. Most skins were exported from Nigeria and Sudan, the countries that imported most of them were Germany, Italy and the

United States. Skins of all three species are also used in the production of stuffed animals which are sold to tourists.

The live animal trade is much smaller than the skin trade and the number of specimens exported were mainly for the European and North American markets. During the period from 1975 to 1986 live exports consisted of only 24 000 *V. salvator*, 13 000 *V. exanthematicus* and 8 000 *V. niloticus*. During that same period 4 233 000, 1 370 000 and 1 210 000 skins of these three species were exported.

Information on population sizes and the status of all these species is very limited. The levels of legislative protection for monitors vary greatly between different countries. Despite this, there is no evident threat to any African species, as they seem to be common in protected or sparsely inhabited areas. Nevertheless, heavy exploitation of African monitors for use by the fashion industry may lead to local extinctions.

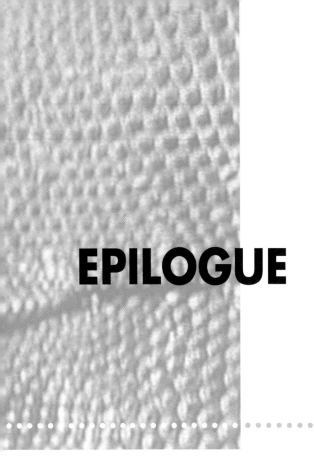

EPILOGUE

She emerged from the burrow, her cold and spent body moving stiffly and slowly onto the damp, cold earth immediately outside. She waited only briefly before moving towards a fallen tree, raising her body upon a small branch and its mat of dead leaves, escaping the cold touch of the ground. She awaited the weak warmth of the winter sun.

The clouds thickened and the wind rose. A shower of rain fell. Still she waited. A heavier shower began; she stirred herself and stolidly returned to the shelter of her burrow, seeking escape from the now pelting rain. She moved to the end of the burrow and curled up in the enlarged chamber there. The little body warmth gained from the winter sun was conserved for a while longer. Thoughts of warmer days, which would allow the insects and small lizards that made up most of her diet to become active, may have entered her brain. If so, they were wishful thoughts. The low winter temperatures would protect them for most of the months ahead as she would lie in the cool, dark burrow, waiting for days when the sun was strong enough to let her emerge and attempt to warm herself above the temperature of the winter soil. Maybe it would be warm enough to enable her to hunt and to restore her depleted body reserves. Maybe. Until then, she slept.

SUGGESTED READING

1. INTRODUCTION

Auffenberg, W. (1981). *The Behavioral Ecology of the Komodo Monitor.* University Presses of Florida: Gainesville, Florida.
—— (1988). *Gray's Monitor Lizard.* University Presses of Florida: Gainesville, Florida.
—— (1994). *The Bengal Monitor.* University Press of Florida: Gainesville, Florida.
Estes, R. (1984). Fish, amphibians and reptiles from the Etadunna formation, Miocene of South Australia. *Australian Zoologist* 21: 335–343.
Hecht, M. K. (1975). The morphology and relationships of the largest known terrestrial lizard, *Megalania prisca* Owen, from the Pleistocene of Australia. *Proceedings of the Royal Society of Victoria* 87: 239–249.
King, M. and King, D. (1975). Chromosomal evolution in the lizard genus *Varanus* (Reptilia). *Australian Journal of Biological Science* 28: 89–108.
Pianka, E. R. (1986). *Ecology and Natural History of Desert Lizards: Analyses of the Ecological Niche and Community Structure.* Princeton University Press, Princeton.
—— (1994). Comparative ecology of *Varanus* in the Great Victoria Desert. *Australian Journal of Ecology* 19: 21–34.

2. TAXONOMY AND PHYLOGENY

Audley-Charles, M. G., Hurley, A. M. and Smith, A. G. (1981). Continental movements in the mesozoic and cenozoic. in *Wallace's Line and Plate*

Tectonics. Clarendon Press, Oxford.

Baverstock, P. R., King, D., King, M., Birrell, J. and Krieg, M. (1993). The evolution of species of the Varanidae: Microcomplement Fixation analysis of serum albumins. *Australian Journal of Zoology* 41: 621–638.

Becker, H. O., Böhme, W. and Perry, S. F. (1989). Die Lungenmorphologie der Warane (Reptilia: Varanidae) und ihre systematisch-stammesgeschichtliche Bedeutung. *Bonn Zoologische Beitrage* 40: 27–56.

Böhme, W. (1988). Zur Genitalmorphologie der Sauria: funktionelle und stammesgeschichtliche Aspekte. *Bonner Zoologische Monographien*, Nr. 27: 1–176.

—— (1995). Hemiclitoris discovered: a fully differentiated erectile structure in female monitor lizards (*Varanus* spp.) (Reptilia: Varanidae). *Journal of Zoological Systematic Evolutionary Research* 33: 129–132.

Branch, W. R. (1982). Hemipeneal morphology of platynotan lizards. *Journal of Herpetology* 16: 16–38.

Carter, D. B. (1990). Courtship and mating in wild *Varanus varius* (Varanidae: Australia). *Memoirs of the Queensland Museum* 29: 333–338.

Clos, L. M. (1995). A new species of *Varanus* (Reptilia: Sauria) from the Miocene of Kenya. *Journal of Vertebrate Paleontology* 15: 254–267.

Estes, R. (1983). Sauriaterrestrie, Amphisbaenia, in *Encyclopaedia of Paleoherpetology*, ed. P. Wellnhofer. G. Fischer Verlag, Stuttgart.

—— (1984). Fish, amphibians and reptiles from the Etadunna formation, Miocene of South Australia. *Australian Zoologist* 21: 335–343.

Fejevary, G. J. (1918). Contributions to a monography on fossil Varanidae and Megalanidae. *Annals Museum National Hungarici* 16: 341–467.

Fuller, S., Baverstock, P. and King, D. (1998). Biogeographic origins of goannas (Varanidae): a molecular perspective. *Molecular Phylogeny and Evolution* 9: 294–307.

Gillespie, J. M., Marshall, R. C. and Woods, E. F. (1982). A comparison of lizard claw keratin proteins with those of avian beak and claw. *Journal of Molecular Evolution* 18: 121–129.

Holmes, R. S., King, M. and King, D. (1975). Phenetic relationships among varanid lizards based upon comparative electrophoretic data and karyotypic analyses. *Biochemical Systematics and Ecology* 3: 257–262.

King, M. (1990). Chromosomal and immunogenetic data; a new perspective on the origin of Australia's reptiles. In *Cytogenetics of amphibians and reptiles: Advances in life sciences*. Birkhauser Verlag, Basel.

King, M and King, D. (1975). Chromosomal evolution in the lizard genus *Varanus* (Reptilia). *Australian Journal of Biological Science* 28: 89–108.

Lee, M. S. Y. (1997). The phylogeny of varanoid lizards and the affinities of snakes. *Philosophical Transactions of the Royal Society, London* B352: 53–91.

Mertens, R. (1942). Die Familie der Warane (Varanidae). Erster Teil: Allgemeines. *Abhandlungen der Senckenbergischen Naturforschenden Gesselenschaft* 465: 1–116.

Pepin, D. J. (in press). A total-evidence phylogeny for the monitor lizards including mtDNA sequence data. *Mertensiella*.

Pianka, E. R. (1995). Evolution of body size: varanid lizards as a model system. *The American Naturalist* 146: 398–414.

Pregill, G. K., Gauthier, J. A. and Greene, H. W. (1986). The evolution of helodermatid squamates, with description of a new taxa and an overview of Varanoidea. *Transactions of the San Diego Natural History Museum* 21: 167–202.

3. FEEDING

Auffenberg, W. (1981). *The Behavioral Ecology of the Komodo Monitor*. University Presses of Florida: Gainesville, Florida.

—— (1988). *Gray's Monitor Lizard*. University Presses of Florida: Gainesville, Florida.

Bellairs, A. d'A. (1949). Observations on the snout of Varanus and a comparison with that of other lizards and snakes. *Journal of Anatomy* 83: 116–146.

Cissé, M. (1972). L'alimentation de Varanidés au Senegal. *Bulletin de l'Institut Francais d'Afrique Noire*, Serie A 34: 503–515.

Cooper, W. E. (1994). Chemical discrimination by tongue-flicking in lizards: a review with hypotheses on its origin and its ecological and phylogenetic relationships. *Journal of Chemical Ecology* 20: 439–487.

James, C. D., Losos, J. B. and King, D. R. (1992). Reproductive biology and diets of goannas (Reptilia: Varanidae) from Australia. *Journal of Herpetology* 26: 128–136.

King, D. and Green, B. (1979). Notes on diet and reproduction of the sand goanna, *Varanus gouldii rosenbergi*. *Copeia* 1979: 64–70.

King, D., Masini, L. and Robinson, R. (in prep). Diet and reproduction of *Varanus scalaris* and *V. tristis*.

Losos, J. B. and Greene, H. W. (1988). Ecological and evolutionary implications of diet in monitor lizards. *Biological Journal of the Linnean Society* 35: 379–407.

Pianka, E. R. (1986). *Ecology and Natural History of Desert Lizards: Analyses of the ecological niche and community structure*. Princeton University Press, Princeton.

Rieppel, O. (1979). A functional interpretation of the varanid dentition (Reptilia, Lacertilia, Varanidae). *Gegenbaurs Morphologie Jahrbuk*, Leipzig 125: 797–817.

Shine, R. (1986). Food habits, habitats and reproductive biology of four sympatric species of varanid lizards in tropical Australia. *Herpetologica* 42: 346–360.

Smith, K. K. (1986). Morphology and function of the tongue and hyoid apparatus in Varanus (Varanidae, Lacertilia). *Journal of Morphology* 187: 261–287.

Traeholt, C. (1994). The food and feeding behavior of the Water monitor, *Varanus salvator*, in Malaysia. *Malayan Nature Journal* 47: 331–343.

Weavers, B. W. (1989). Diet of the Lace Monitor Lizard (*Varanus varius*) in south-eastern Australia. *Australian Zoologist* 25: 83–85.

4. BREEDING

Auffenberg, W. (1981). *The Behavioral Ecology of the Komodo Monitor*. University Presses of Florida: Gainesville, Florida.

—— (1994). *The Bengal Monitor*. University Press of Florida: Gainesville, Florida.

Bredl, J. and Horn, H.-G. (1987). Über die Nachzucht des australischen Riesewarans *Varanus giganteus* (Gray, 1845). *Salamandra* 23: 90–96.

Cowles, R. B. (1930). The life history of *Varanus niloticus* (Linnaeus) as observed in Natal, South Africa. *Journal of Entomology and Zoology* 22: 1–31.

Carter, D. B. (1990). Courtship and mating in wild *Varanus varius* (Varanidae: Australia). *Memoirs of the Queensland Museum* 29: 333–338.

Darevsky, I. S. and Auffenberg, W. (1973). Growth records of wild recaptured Komodo monitors (*Varanus komodoensis*). *HISS News-Journal* 1: 41–42.

Horn, H.-G. and Visser G. (1989). Review of reproduction of monitor lizards (*Varanus* spp.) in captivity. *International Zoo Yearbook* 29: 140–150.

—— (1991). Basic data on the biology of monitors. *Mertensiella* 2: 176–187.

—— (1997). Review of reproduction of monitor lizards (*Varanus* spp.) in captivity II. *International Zoo Yearbook* 35: 227–246.

Irwin, I., Engle, K. and Mackness, B. (1996). Nocturnal nesting by captive varanid lizards. *Herpetological Review* 27: 192–194.

King, D. and Green, B. (1979). Notes on diet and reproduction of the sand goanna, *Varanus gouldii rosenbergi*. *Copeia* 1979: 64–70.

King, D., Green, B. and Butler, H. (1989). The activity pattern, temperature regulation and diet of *Varanus giganteus* on Barrow Island, Western Australia. *Australian Wildlife Research* 16: 41–47.

King, D., Masini, L. and Robinson, R. (in prep.). Diet and reproduction of *Varanus scalaris* and *V. tristis*.

King, D. and Rhodes, L. (1982). Sex ratio and breeding season of *Varanus acanthurus*. *Copeia* 1982: 784–787.

Phillips, J. A. (1995). Movement patterns and density of *Varanus albigularis*. *Journal of Herpetology* 29: 339–348.

—— (In press). Reproductive biology of the White-throated Savanna monitor, *Varanus albigularis*.

Riley, J., Stimson, A. F. and Winch, J. M. (1985). A review of squamata ovipositing in ant and termite nests. *Herpetological Review* 16: 38–43.

Rookmaaker, L. C. (1975). The history of some Komodo dragons (*Varanus komodoensis*) captured on Rintja in 1927. *Zoologische Mededelingen* 49: 65–71.

Shine, R. (1986). Food habits, habitats and reproductive biology of four sympatric species of varanid lizards in tropical Australia. *Herpetologica* 42: 346–360.

5. GENERAL BEHAVIOUR

Cowles, R. B. (1930). The life history of *Varanus niloticus* (Linnaeus) as observed in Natal, South Africa. *Journal of Entomology and Zoology* 22: 1–31.

Green, B. (1972). Water losses of the sand goanna (*Varanus gouldii*) in its natural environment. *Ecology* 53: 452–457.

Green, B. and King, D. (1978). Home range and activity patterns of the sand goanna, *Varanus gouldii* (Reptilia: Varanidae). *Australian Wildlife Research* 5: 417–424.

Horn, H.-G. (1985). Beitrage zum Verhalten von Waranen: die Ritualkampfe von *Varanus komodoensis* Ouwens, 1912 und *V. semiremex* Peters, 1869 sowie die Imponierphasen der Ritualkampfe von *V. timorensis timorensis* (Gray, 1931) und *V. t. similis* Mertens, 1958. *Salamandra* 21: 169–179.

Horn, H.-G., Gaulke, M. and Böhme, W. (1994). New data on ritualized combats in monitor lizards (Sauria: Varanidae), with remarks on their function and phylogenetic implications. *Der Zoologische Garten* 64: 265–280.

James, C. D. (1996). Ecology of the pygmy goanna (*Varanus brevicauda*) in spinifex grasslands of central Australia. *Australian Journal of Zoology* 44: 177–192.

Murphy, J. B. and Mitchell, L. A. (1974). Ritualized combat behavior of the pygmy mulga monitor lizard, *Varanus gilleni* (Sauria: Varanidae). *Herpetologica* 30: 90–97.

Pianka, E. R. (1986). *Ecology and Natural History of Desert Lizards: Analyses of the ecological niche and community structure.* Princeton University Press, Princeton.

Regal, P. J. (1978). Behavioral differences between reptiles and mammals: an analysis of activity and mental capabilities. In *Behavior and neurology of lizards*, eds N. Greenberg and P.J. MacLean, National Institute of Mental Health (NIMH), Rockville, Maryland.

Shine, R. (1986). Food habits, habitats and reproductive biology of four sympatric species of varanid lizards in tropical Australia. *Herpetologica* 42: 346–360.

Stanner, M. and Mendelssohn, M. (1987). Sex ratio, population density and home range of the desert monitor (*Varanus griseus*) in the southern coastal plain of Israel. *Amphibia–Reptilia* 8: 153–164.

Vernet, R., Lemire, M., Grenot, C. J. and Francaz, J. M. (1988). Ecophysiological comparisons between two large Saharan lizards, *Uromastix acanthinurus* (Agamidae) and *Varanus griseus* (Varanidae). *Journal of Arid Environments* 14: 187–200.

—— (1988). Field studies on activity and water balance of a desert monitor *Varanus griseus* (Reptilia, Varanidae). *Journal of Arid Environments* 15: 81–90.

6. THERMAL BIOLOGY

Bowker, R. G. (1984). Precision of thermoregulation of some African lizards. *Physiological Zoology* 57: 401–412.

Christian, K. and Bedford, G. (1996). Thermoregulation by the spotted tree monitor, *Varanus scalaris*, in the seasonal tropics of Australia. *Journal of Thermal Biology* 21: 67–73.

Christian, K., Bedford, G. S. and Shannahan, S. D. (1996). Solar absorptance of some Australian lizards and its relationship to temperature. *Australian Journal of Zoology* 44: 59–67.

Cowles, R. B. and Bogert, C. M. (1944). A preliminary study of the thermal requirements of desert reptiles. *Bulletin of the American Museum of Natural History* 83: 261–296.

Green, B., King, D., Braysher, M. and Saim, A. (1991). Thermoregulation, water turnover and energetics of free-living Komodo dragons, *Varanus*

komodoensis. Comparative Biochemistry and Physiology 99A: 97–101.
Huey, R. B. (1974). Behavioural thermoregulation in lizards: importance of associated costs. *Science* 184: 1001–1003.
Hutchinson, V. H. and Larimer, J. L. (1960). Reflectivity of the integuments of some lizards from different habitats. *Ecology* 41: 199–209.
King, D. (1980). The thermal biology of free-living sand goannas (*Varanus gouldii*) in southern Australia. *Copeia* 1980: 755–767.
King, D., Green, B. and Butler, H. (1989). The activity pattern, temperature regulation and diet of *Varanus giganteus* on Barrow Island, Western Australia. *Australian Wildlife Research* 16: 41–47.
Murrish, D. E. and Schmidt-Nielsen, K. (1972). Exhaled air temperature and water conservation in lizards. *Respiration Physiology* 10: 151–158.
Sokholov, V.E., Sukhov, V.P. and Chernyshov, Y. M. (1975). A radio telemetric study of diurnal fluctuations of body temperature in the desert monitor (*Varanus griseus*). *Zoological Zhurnal* 54: 1347–1356.
Traeholt, C. (1995). A radio telemetric study of the thermoregulation of free-living water monitor lizards, *Varanus salvator*. *Journal of Comparative Physiology* 165: 125–131.
Vernet, R., Lemire, M., Grenot, C. J. and Francaz, J. M. (1988). Ecophysiological comparisons between two large Saharan lizards, *Uromastix acanthinurus* (Agamidae) and *Varanus griseus* (Varanidae). *Journal of Arid Environments* 14: 187–200.
Weavers, B. W. (1983). Thermal ecology of *Varanus varius* (Shaw), the Lace Monitor. Unpublished Ph.D. thesis, Australian National University, Canberra.
Wikramanayake, E. and Green, B. (1989). Thermoregulatory influences on the ecology of two sympatric varanids in Sri Lanka. *Biotropica* 21: 74–79.

7. RESPIRATION

Becker, H. O., Böhme, W. and Perry, S. F. (1989). Die Lungenmorphologie der Warane (Reptilia: Varanidae) und ihre systematisch-stammesorschichtliche Bedeteung. *Bonn Zoologische Beitrage* 40: 27–56.
Bellairs, A. d'A. (1949) Observations on the snout of Varanus, and a comparison with that of other lizards and snakes. *Journal of Anatomy* 83: 116–146.
Bennett, A. F. (1972). The effect of activity on oxygen consumption, oxygen debt, and heart rate in the lizards *Varanus gouldii* and *Sauromalus hispidus*. *Journal of Comparative Physiology* 79: 259–280.
—— (1973). Blood physiology and oxygen transport during activity in two lizards, *Varanus gouldii* and *Sauromalus hispidus*. *Comparative Biochemistry and Physiology* 46A: 673–690.
Glass, M. L., Wood, S. C., Hoyt, R. W. and Johansen, K. (1979). Chemical control of breathing in the lizard, *Varanus exanthematicus*. *Comparative Biochemistry and Physiology* 62A: 999–1003.
Heisler, N., Neumann, P. and Maloiy, G. M. O. (1983). The mechanism of intracardiac shunting in the lizard *Varanus exanthematicus*. *Journal of Experimental Biology* 105: 15–31.

Perry, S. F. (1983). *Reptilian lungs. Advances in anatomy, embryology and cell biology*. Springer Verlag, Berlin, Heidelberg.

Webb, G., Heatwole, H. and DeBavay, J. (1971). Comparative cardiac anatomy of the reptilia 1. The chambers and septa of the varanid ventricle. *Journal of Morphology* 134: 335–350.

Wood, S. C. and Johansen, K. (1974). Respiratory adaptations to diving in the Nile Monitor Lizard, *Varanus niloticus*. *Journal of Comparative Physiology* 89: 145–158.

8. WATER USE

Braysher, M. and Green, B. (1970). Absorption of water and electrolytes from the cloaca of an Australian lizard, *Varanus gouldii* (Gray). *Comparative Biochemistry and Physiology* 35: 607–614.

Christian, K. A., Weavers, B. W., Green, B. and Bedford, G. (1996). Energetics and water flux in a semi-aquatic lizard, *Varanus mertensi*. *Copeia* 1996: 354–362.

Dryden, G., Green, B., King, D. and Losos, J. (1990). Water and energy turnover in a small monitor lizard, *Varanus acanthurus*. *Australian Wildlife Research* 17: 641–646.

Green, B. (1972). Aspects of renal function in the lizard *Varanus gouldii*. *Comparative Biochemistry and Physiology* 43A: 747–756.

—— (1972) Water losses of the sand goanna (*Varanus gouldii*) in its natural environment. *Ecology* 53: 452–457.

Green, B., King, D. and Butler, W. H. (1986). Water, sodium and energy metabolism in free-living Perenties, *Varanus giganteus*. *Australian Wildlife Research* 34: 589–595.

Green, B., King, D., Braysher, M. and Saim, A. (1991). Thermoregulation, water turnover and energetics of free-living Komodo dragons, *Varanus komodoensis*. *Comparative Biochemistry and Physiology* 99A: 97–101.

Vernet, R., Lemire, M. and Grenot, C. (1988). Field studies on activity and water balance of a desert monitor *Varanus griseus* (Reptilia, Varanidae). *Journal of Arid Environments* 15: 81–90.

9. ENERGY AND FOOD

Christian, K. A., Corbett, L. K., Green, B. and Weavers, B. W. (1995). Seasonal activity and energetics of two species of varanid lizards in tropical Australia. *Oecologia* 103: 349–357.

Christian, K., Green, B., Bedford, G. and Newgrain, K. (1996). Seasonal metabolism of a small, arboreal monitor lizard, *Varanus scalaris*, in tropical Australia. *Journal of Zoology, London* 240: 383–396.

Dryden, G., Green, B., King, D. and Losos, J. (1990). Water and energy turnover in a small monitor lizard, *Varanus acanthurus*. *Australian Wildlife Research* 17: 641–646.

Dryden, G., Green, B., Wikramanayake, E. and Dryden, K. (1992). Energy and water turnover in two tropical varanid lizards, *Varanus bengalensis* and *V. salvator*. *Copeia* 1992: 102–107.

Gales, R. and Green, B. (1990). The annual energetics of Little Penguins (*Eudyptula minor*). *Ecology* 71: 2297–2312.

Green, B., Dryden, G. and Dryden, K. (1991). Field energetics of a large carnivorous lizard, *Varanus rosenbergi*. *Oecologia* 88: 547–551.

Green, B., King, D., Braysher, M. and Saim, A. (1991). Thermoregulation, water turnover and energetics of free-living Komodo dragons, *Varanus komodoensis*. *Comparative Biochemistry and Physiology* 99A: 97–101.

Green, B., King, D. and Butler, W. H. (1986). Water, sodium and energy metabolism in free-living Perenties, *Varanus giganteus*. *Australian Wildlife Research* 34: 589–595.

Lifson, N. and McClintock, R. (1966). Theory of use of the turnover rates of body water for measuring energy and material balance. *Journal of Theoretical Biology* 12: 46–74.

Nagy, K. A. (1982). Energy requirements of free-living iguanid lizards. In *Iguanas of the World: Their behavior, ecology and conservation*. Noyes Publications, Park Ridge, New Jersey.

Vernet, R., Lemire, M., Grenot, C. J. and Francaz, J. M. (1988). Ecophysiological comparisons between two large Saharan lizards, *Uromastix acanthinurus* (Agamidae) and *Varanus griseus* (Varanidae). *Journal of Arid Environments* 14: 187–200.

10. PARASITES

Auffenberg, T. (1988). *Amblyomma helvolum* (Acarina: Ixodidae) as a parasite of varanid and scincid reptiles in the Philippines. *International Journal of Parasitology* 18: 937–945.

Auffenberg, W. (1981). *The Behavioral Ecology of the Komodo Monitor*. University Presses of Florida: Gainesville, Florida.

Auffenberg, W. and Auffenberg, T. (1990). The reptile tick *Aponomma gervaisi* (Acarina: Ixodidae) as a parasite of monitor lizards in Pakistan and India. *Bulletin of the Florida Museum of Natural History, Biological Sciences* 35: 1–34.

Bosch, H. (1991). The pentastome *Elenia* Heymons, 1932, a parasite genus specific to monitors from the Australian region – biology and aspects of host specificity. *Mertensiella* 2: 50–56.

Bull, C. M. and Burzacott, D. (1993). The impact of tick load on the fitness of their lizard hosts. *Oecologia* 96: 415–419.

Chilton, N. B. (1994). Differences in the life cycles of two species of reptile ticks: implications for species distributions. *International Journal for Parasitology* 24: 791–795.

Jones, H. I. (1983). Prevalence and intensity of *Abbreviata* Travassos (Nematoda: Physalopteridae) in the Ridge-tailed monitor *Varanus acanthurus* Boulenger in northern Australia. *Records of the Western Australian Museum* 11: 1–9.

—— (1988). Nematodes from nine species of *Varanus* (Reptilia) from tropical northern Australia with particular reference to the genus *Abbreviata* (Physalopteridae). *Australian Journal of Zoology* 36: 691–708.

—— (1991). Speciation, distribution and host-specificity of gastric Nematodes in Australian Varanid lizards. *Mertensiella* 2: 195–203.

—— (1995). Gastric nematode communities in lizards from the Great

Victoria Desert, and an hypothesis for their evolution. *Australian Journal of Zoology* 43: 141–164.

Keirans, J. E., King, D. R. and Sharrad, R. D. (1994). *Aponomma (Bothriocroton) glebopalma*, n. subgen., n. sp., and *Amblyomma glauerti* n. sp. (Acari: Ixodida: Ixodidae), parasites of monitor lizards (Varanidae) in Australia. *Journal of Medical Entomology* 31: 132–147.

King, D. R. and Keirans, J. E. (1997). Ticks (Acari: Ixodidae) from varanid lizards in eastern Indonesia. *Records of the Western Australian Museum* 18: 329–330.

Lenz, S. (1995). Zur biologie und okologie des Nilwarans *Varanus niloticus* (Linnaeus 1776) in Gambia, Westafrika. *Mertensiella* 5: 12–256.

Roberts, F. H. S. (1970). *Australian ticks*. CSIRO, Melbourne.

Sharrad, R. D. and King, D. R. (1981). The geographical distribution of reptile ticks in Western Australia. *Australian Journal of Zoology* 29: 861–873.

Ward, D. L. (1989). Subdermal infestation of a monitor lizard by *Aponomma undatum* (Fabricius) (Acarina: Ixodidae). *Australian Entomology Magazine* 16: 78–79.

11. CONSERVATION AND MANAGEMENT

Auffenberg, W. (1981). *The Behavioral Ecology of the Komodo Monitor*. University Presses of Florida: Gainesville, Florida.

de Buffrenil, V. (1992). La peche et l'exploitation du varan Nil (*Varanus niloticus*) dans la région du lac Tchad. *Bulletin de la Societé Herpelotogique de France* 62: 47–56.

Erdelen W. (1991). Conservation and population ecology of monitor lizards: the water monitor *Varanus salvator* (Laurenti, 1768) in south Sumatra. *Mertensiella* 2: 120–135.

Gould, R. A. (1980). *Living Archaeology*. Cambridge University Press, Cambridge.

Hoogwerf, A. (1970). Common Monitor Lizard, in *Udjung Kulon. The Land of the Last Javan Rhinoceros*. Leiden.

Luxmoore, R., Groombridge, B. and Broad, S. (eds). (1988). Significant trade in wildlife: a review of selected species, in *CITES Appendix II. Reptiles and invertebrates*. International Union for the Conservation of Nature and Natural Resources, Gland, Switzerland.

Shine, R., Harlow, P. S. and Keogh, J. S. (1996). Commercial harvesting of giant lizards: The biology of water monitors *Varanus salvator* in southern Sumatra. *Biological Conservation* 77: 125–134.

Vernet, R. and Chamaillard, M. (1984). Les varans. In *Especes menacées et exploitées dans le monde; guide pratique pour leur connaissance et leur identification*. ed. F. de Beaufort, Secretariat de la Faune et la Flore, Paris.

INDEX

Abbreviata 92, 93
Aboriginal culture, goannas in 96
activity
　areas 39–41
　levels 19, 39, 42, 43, 62
　patterns 42
　period 42, 52, 57
aerobic capacity 62
albumin 10
Amblyomma 89, 90
ambush 18
Amphibolurus maculosus 71
Aponomma 89, 90, 92, 94
arboreal species 6, 16, 19, 40, 54–56, 66

basking 27, 49, 52–54, 56, 57, 70
beaded lizards 6, 54
bias, male 42
bite 23, 29, 30
body size 34, 35, 55, 69, 79, 87
breeding 59, 85, 99
　activity areas 40
　egg-laying 31–33
　hatching 35–36
　mating 27–30
　sex organs 25–26

Bufo marinus 96
burrows 3, 40, 42, 44, 50–52
　body temperature 49, 56–57, 58
　dehydration 68
　egg-laying 33
　feeding 19
　mating 27, 28
　water turnover 78

cannibalism 18, 36
carnivorous species 5, 7, 16, 24, 40,
　66, 76, 79, 81, 82, 87, 100
carrion 17, 18, 85
chromosome 10, 12–14
cloaca 6, 24, 25, 27, 32, 56, 71,
　73–76, 90, 92
clutch 29, 31, 34, 35, 85
　size 34, 85
colour, skin 6, 59, 98
conservation 95, 97, 100
coprodaeum 73–75
copulation 25, 27, 29
cranial kinesis 24

deuterium 77, 82
diaphragm 61

INDEX

diet 5, 6, 16–19, 82, 84, 92, 94, 95
distribution 6–8, 13, 49, 90, 94, 96, 98
drinking 23, 66, 77–79

earless monitor 6, 7
ectoparasite 89
efflux 77
eggs 33
  chamber 31, 32, 36
  development 34
  egg-laying 31–34, 36
  in diet 7, 17–19, 43
  parasites 90, 92, 93
  reptiles 33
  varanids 28, 31–35, 84, 85, 97
electrophoretic studies 10
embrace 29
endoparasite 89, 92, 94
energy budget 84, 85
excavation 31, 95
exploitation 96–99, 101
exportation of goannas 97, 98, 100, 101
eyes 6, 20, 57, 58, 69, 70, 75, 76

fat bodies 86, 97
food 16–20, 30, 81–88, 93–98, 100
  for breeding 35
  carnivores 5, 7, 16, 24, 40, 66, 76, 79, 81, 82, 87–88, 100
  detection 20
  eggs 31
  fat storage 28
  foraging 43
  growth rates 37
  handling 21–24
  herbivores 16, 24, 40, 66, 76, 82, 87–88
  insectivores 5, 41, 81
foraging 17, 18, 20
  after copulation 27
  by hatchlings 36
  range 39–40
  strategies 43, 65
fossils 8, 9, 12

gene-sequencing 12, 14
Goanna Oil 97
Gondwana 12, 13
gravid 34
growth rate 37, 92
gular fluttering 58, 59, 67, 68, 70
gular pouch 46, 67
gular region 6, 23, 58

habitats 6, 19, 43, 59, 66, 99
  alterations 96

burrows 49–50
  destruction of 99
  energy 87
  parasites 90, 92–94
  water use 79–80, 87
hatchlings 28, 35–37, 46, 69, 86
heart 61, 64, 65
Heloderma 7, 54, 61
helodermatids 6, 7, 9
hemibaculum 25
hemipenes 13, 14, 25, 27, 29
herbivores 16, 24, 40, 66, 76, 82, 87
herbivorous 16, 24, 40, 76
home range 29, 39–41, 43
humidity 33, 51, 67, 79
hyoid apparatus 21–24

importation, of goannas 98, 100
incubation 28, 34, 35
  chamber 33
  period 34, 35
inertial feeding 23, 24
influx 77, 78, 84
insectivores 5, 41, 81
invertebrates 6, 7, 17–20, 92, 93
isotope 77, 82, 83

Jacobson's organs 17, 20–22
jaw 3, 23, 24

kidneys 71–73, 75, 76

lanthanotids 7
Lanthanotus 6, 7, 61
legislation 95, 99, 101
length
  snout–vent 37
  tongue 22
  total body 6, 7, 36, 37
lipids 31, 68
longevity 37, 92
lungs 61–65, 67, 70, 77, 79, 94

mating 25, 27–29, 32, 33, 92
mating assemblages 29
MC'F 10, 12
Megalania prisca 6, 7
metabolic rate 67, 79, 81, 82–84, 86, 88
micro–complement fixation 11
morphology 10, 14, 39
  chromosome 12
  eye 6
  kidney 71
  lung 13, 63
  snout 20
mosasaurs 7

nares 20, 58, 76, 90
nasal capsule 20, 75
nasal gland 76
nasal sac 20, 21
nasal tubes 20, 58
Necrosauridae 9
nematodes 89, 92–94
nest 32, 33, 35, 36, 43

olfactory chamber 20, 21, 76
osteology 10
ova 26
ovaries 25, 26
oviducts 26, 31

pair bond 29
phylogeny 10, 12
platynotans 9
postures 45–47
predators 18, 34, 98
  aggressive postures 46
  burrows 44
  diurnal 32
  eggs 33
  hatchlings 36
prey 6, 7, 16–20, 43, 92, 93, 96, 100
  amount eaten 87–88
  availability of 78
  digestion of 82
  energy content of 84–85
  large 17, 23, 24
proctodaeum 73

radio-telemetry 53, 54
reflectivity 59, 60
respiration 56, 61–64, 67, 81
  aerobic 62–64
  anaerobic 62, 64
ritual combat 29, 30, 46, plate 21

salt-secreting glands 20, 75–76
semi-aquatic species 6, 19, 49, 53, 54, 56, 59, 63, 65, 66
sex organs 25–26
sex ratio 42, 43
sexual maturity 37
skin 20, 59, 60, 67–69, 70, 77, 79, 90, 96, 99
skins, harvested 95, 97–101
snakes 7, 9
  evaporation 70
  harvesting for skins 99
  JacobsonÕs organs 20
  parasites 89, 92
  postures against 46

salt glands 75
  tongue 21
snout 17, 20, 23, 58
solar radiation 54
survival rates 37

taxonomic studies 8, 14
taxonomy 10
teeth 3, 17, 23, 24, 30
Telmasaurus 9
temperature 33, 63
  activity 49, 53–56, 60. 62
  air 32, 50, 51, 54, 59, 60, 67, 69, 102
  body 19, 40, 49, 50–59, 62, 64, 65, 67, 68, 72, 86, 87–88
  burrow 51, 52
  gradient 54, 56, 58
  head 52, 56–59
  incubation 28, 33–35
  regulation 53, 54
  soil 42, 50–52, 102
  surface 51, 54
termitarium 31–33, 35, 36
termite mounds 31–33, 50, 86, plate 8
termites 32, 33, 92, 93
testes 25, 26
ticks 89–92, 94
Tiliqua rugosa 92
tongue 3, 17, 21–23, 76, 94, plate 7
traditional medicines 97, 100
tritium 77, 82

upright stance 29, 48, plate 7
urinary pellet 75, 76
urine 67, 71–75, 81
urodaeum 73

varanoids 1, 9
Varanus
  acanthurus 11, 12, 15, 17, 18, 34, 35, 79, 86, 91
  albigularis 11, 12, 15, 100
    activity patterns 42
    breeding 26, 28, 29
    captivity 38
    clutch size 34
    egg-laying 31, 33
    feeding 19
    foraging strategies 43
    hatching 36
    home range 39, 40
  baritji 11, 15, 17
  bengalensis 6, 11, 12, 15, 90, 97
    activity patterns 42

INDEX

body temperature 50, 56
breeding 28–30, 33, 35–38
feeding 18, 19
metabolic rate 86
parasites 90, 91, 94
water turnover 79
brevicauda 6, 7, 11, 12, 15, 17, 18, 26, 34, 35, 40
caudolineatus 6, 11, 15, 19, 29, 40, 42, 50, 56, 79, 86, 91
doreanus 15
dumerilii 11, 12, 15, 29, 35, 36, plate 16
eremius 11, 12, 15, 19, 35, 42, 50, 56, 91
exanthematicus 11, 15, 31, 38, 56, 63, 100, 101
finschi 15
flavescens 11, 12, 15, 38, 75, 90, 97, 99
giganteus 6, 11, 12, 15, plate 6
   activity patterns 42
   activity temperatures 54, 55, 56
   burrows 50, 52
   clutch size 35, 39
   egg-laying 31, 33
   feeding 17–19
   home range 40
   metabolic rate 86
   parasites 91, 94
   postures 46
   water turnover 79
gilleni 11, 12, 14, 15
   breeding 29, 30, 31
   burrows 50
   feeding 17
   parasites 91
   temperatures 56
   water use 68
glauerti 6, 11, 12, 15, 17, 19, 36, 42, 91, plate 12, plate 14
glebopalma 11, 15, 17, 18, 42, 91
gouldii 6, 8, 11, 12, 14, 15, plate 20
   activity patterns 42
   activity temperatures 54, 55, 56
   AVT 72
   breeding 29
   burrows 44, 50, 51, 52
   captivity 38
   egg-laying 31, 33, 35
   feeding 18, 19, 23
   gular fluttering 59
   home range 39
   kidney function 73
   metabolic rate 86

   parasites 90, 94
   reflectivity of the skin 60
   water loss 68
   water turnover 79
griseus 8, 11, 12, 15, 99, 100
   activity temperatures 54, 55, 56
   breeding 28, 33
   burrows 50, 52
   captivity 38
   feeding 18, 19
   home range 39
   parasites 90, 91
   salt-secreting glands 75
   water turnover 79
indicus 11, 12, 15, 97, 98
   activity patterns 42
   breeding 29, 31, 35
   feeding 18, 19,
   parasites 91, 93
   water use 66
jobiensis 11, 12, 15, 66
kingorum 11, 12, 15, 96
komodoensis 6, 11, 12, 14, 15, plates 2 and 4
   activity patterns 42
   activity temperatures 54, 55, 56
   breeding, 26, 28–31
   burrows 50, 52
   captivity 38
   egg-laying 33
   feeding 18–20, 23
   foraging strategies 43
   growth rate 37
   gular fluttering 59
   hatching 35, 36
   longevity 39
   movement 46
   parasites 91, 93, 94
   water turnover 79
melinus 15
mertensi 11, 12, 15, , plates 5 and 21
   activity patterns 42
   activity temperatures 54, 56
   egg-laying 31
   feeding 19, 20
   mating 29
   metabolic rate 86
   parasites 93
   water use 66, 79
mitchelli 11, 12, 15, 42, 66, 91
niloticus 11, 12, 15, 100, 101
   activity temperatures 54, 56
   breeding 29, 30, 33, 36, 38
   feeding 18, 19
   foraging 43

parasites 91
  water use 66
olivaceus 5, 11, 15, 16, 28–30, 91
ornatus 15
panoptes 11, 12, 15, plate 7
  activity patterns 42
  breeding 29, 33
  burrows 44, 50
  feeding 19, 23
  metabolic rate 86
  parasites 91
  postures 48
  water use 79
'pellewensis' plate 3
pilbarensis 11, 15, 36, 91, 96, plate 18
prasinus 11, 12, 15, 31, 35, 36, 94, plate 19
primordius 11, 12, 14, 15
rosenbergi 7, 8, 11, 12, 15, plates 1 and 9–11
  basking 52
  breeding 26, 27, 28, 29, 31, 32, 33, 36, 37
  burrows 44, 50–51
  energy consumption 83, 84, 85, 86, 87, 88
  feeding 17, 18, 24
  home range 39–42
  parasites 91
  temperatures 50–60
  water use 67–70, 72–75, 77–80, 87–88
rudicollis 11, 12, 15, 35
salvadorii 6, 11, 15, 35, 36
salvator 6, 11, 12, 15, 97–99, 101, plate 13
  activity patterns 42
  breeding 26, 29, 30, 31, 33–36
  burrows 50
  captivity 38
  feeding 18, 19
  metabolic rates 86
  parasites 91, 94
  temperatures 53–56, 59
  water use 66, 75, 79
scalaris 11, 12, 15
  activity patterns 42
  breeding 34
  burrows 50
  feeding 19

  metabolic rate 86
  parasites 91, 93
  temperatures 54, 55
  water use 79
semiremex 11, 15, 29, 42, 66, 75
similis 15, 29
spenceri 11, 12, 15, 29, 35
spinulosus 15
storri 11, 15, 36, 42, 68, plate 15
timorensis 11, 15, 19, 29, 35, 91, 93
tristis 6, 11, 12, 15, plate 17
  activity patterns 42
  breeding 28, 29, 31, 33–36
  burrows 50
  feeding 18, 19
  home range 39, 40
  parasites 91, 94
  temperatures 54–56
  water use 75
varius 8, 11, 12, 15, 97
  activity patterns 42
  breeding 28, 29, 30, 31, 33, 35–37
  burrows 44, 50, 52
  feeding 17, 18, 19,
  home range 39
  parasites 90, 94
  temperatures 54–57
  water use 77, 79
yemenensis 11, 15
yuwonoi 15
venomous 7, 28, 46
vertebrates 17, 19, 20, 68, 73, 87, 89, 96

water
  balance 66, 72, 77, 78
  conservation 58, 75
  intake 77, 78, 83, 87–88
  isotopic 77
  loss 51, 58, 67, 68, 77, 82
    cutaneous 68, 69
    evaporative 51, 58, 67–70
    excretory 75, 78
    respiratory 70
  turnover 77–79, 80, 82, 87
weight
  body 6, 20, 31, 37, 40, 61, 69, 79, 84, 86, 87
  egg 33
wide-foraging 18